Geology of the country around Cadair Idris

The Cadair Idris district, in the south of the Snowdonia National Park, comprises Cambrian to Silurian sedimentary strata with extrusive volcanic rocks and igneous intrusions. Ordovician volcanic strata form the highest ground, including the summit of Cadair Idris, whereas the sedimentary rocks, form lower, gently rounded and deeply dissected terrain.

The oldest Cambrian sedimentary strata are mainly turbiditic sandstones and mudstones, with repeated beds of laminated, pelagic black mudstone. The pattern of Cambrian basinal sedimentation was broken by uplift and subaerial Tremadoc volcanic activity, which produced the Rhobell Volcanic Group, a sequence of intermediate lavas and tuffs. Erosion removed up to 1 km of Cambrian and Tremadoc strata prior to a marine transgression in Arenig times.

Subsequently, rapid subsidence occurred, with deposition of mainly black mudstones during the emplacement of the Aran Volcanic Group (Arenig–Caradoc). The sources of the volcanic rocks lay outside the district. The extensive development of pillowed basalts within the sequence indicates subaqueous accumulation. Many of the acid and basic volcanic deposits were remobilised into muddy debris-flow deposits. The final phase of volcanism occurred with the voluminous eruption and emplacement of acid ash-flow tuffs. Later Ordovician and Silurian deposition was dominated by turbiditic, submarine fan deposition.

The climactic end-Silurian (Caledonian) orogeny folded and cleaved the Lower Palaeozoic strata and metamorphosed them to low greenschist grade. Later, Variscan (late Carboniferous) deformation was probably restricted to fault movements, in particular strike-slip movement on the Bala and Tal-y-llyn faults. Mesozoic activity along major coastal faults caused the development of offshore sedimentary basins which are currently being explored for oil and gas. Glaciation during the Quaternary produced many characteristic landforms, including cwms, glacial lakes, moraines and U-shaped valleys.

Cover photograph

The summit of Cadair Idris, Penygadair, viewed from Mynydd Moel [7280 1360].

Looking south-west towards Tal-y-llyn and the landslip scar (mainly in shadow).

BRITISH GEOLOGICAL SURVEY

W T PRATT
D G WOODHALL and
M F HOWELLS

Geology of the country around Cadair Idris

Memoir for 1:50 000 geological sheet 149
(England and Wales)

CONTRIBUTORS

Ashgill
M J Leng

Silurian
R Cave
M S Scott

Geophysics
R M Carruthers

Palaeontology
S G Molyneux
A W A Rushton
J A Zalasiewicz

LONDON: HMSO 1995

First published 1995

ISBN 0 11 884509 8

Bibliographical reference

Pratt, W T, Woodhall D G, and Howells, M F. 1995. Geology of the country around Cadair Idris. *Memoir of the British Geological Survey,* Sheet 149 (England and Wales).

Authors

W T Pratt, BSc, PhD
British Geological Survey
Keyworth

D G Woodhall, BSc, PhD
British Geological Survey
Edinburgh

M F Howells, BSc, PhD
University of Liverpool

Contributors

R M Carruthers, BSc, PhD
S G Molyneux, BSc, PhD
A W A Rushton, BA, PhD
British Geological Survey
Keyworth

J A Zalasiewicz, BSc, PhD
University of Leicester

R Cave, BSc, PhD
University of Wales, Aberystwyth

M J Leng, BSc, PhD
NERC Isotope Laboratories
Keyworth

M S Scott, BSc, PhD
Department of the Environment
London

Printed in the UK for HMSO

Dd 296602 C8 3/95

Other publications of the Survey dealing with this and adjoining districts

BOOKS

British regional geology
North Wales 3rd edition, 1961, reprinted 1987
South Wales 3rd edition, 1970, reprinted 1982

Memoirs
Geology of the country around Harlech (Sheet 135), 1985
Geology of the country around Aberystwyth (Sheet 163), 1986

Mineral Reconnaissance Programme reports
No. 5 Preliminary mineral reconnaissance of Central Wales
No. 74 A reconnaissance geochemical drainage survey of the Harlech Dome, North Wales

Field guide
Geological excursions in the Harlech Dome, 1985

Special memoir
Ordovician (Caradoc) marginal basin volcanism in Snowdonia (north-west Wales), 1990

MAPS

1:1 000 000
Geology of the United Kingdom, Ireland and adjacent continental shelf (South Sheet), 1991

1:625 000
Solid geology (South Sheet), 1979
Quaternary geology (South Sheet), 1977
Aeromagnetic map (South Sheet), 1965

1:250 000
Solid geology, Cardigan Bay, 1982
Sea bed sediments, Cardigan Bay, 1988
Quaternary, Cardigan Bay, 1990
Aeromagnetic anomaly, Cardigan Bay, 1980
Bouguer gravity anomaly, Cardigan Bay, 1980
Solid geology, Mid Wales and the Marches, 1990
Aeromagnetic anomaly, Mid Wales and the Marches, 1980
Bouguer gravity anomaly, Mid Wales and the Marches, 1986

1:100 000
Central Wales mining field, 1974

1:50 000
Sheet 135 Harlech (Solid and Drift) 1982
Sheet 136 Bala (Solid) 1986
Sheet 163 Aberystwyth (Solid) 1984, (Drift) 1989

CONTENTS

FIGURES

PLATES

TABLES

PREFACE

The Cadair Idris district lies at the southern edge of the Snowdonia National Park and, from the high ridges of the Cadair Idris massif to the tidal reaches of the Mawddach and Dyfi estuaries, it is an area of spectacular beauty. For many geologists it is a special place as it lies close to the heart of Sedgwick's Cambrian System. The survey of the Cadair Idris district has followed those of the Harlech, Snowdon and Bangor districts to the north and has broadened the correlation of the Cambrian and Ordovician sequences in north-west Wales. In particular, the study of the Ordovician volcanic rocks allows comparison to be made with recent major studies in central and northern Snowdonia. The mapping of the Silurian sequence also allows it to be considered in context with the broad swathe of similar strata recently mapped on the Aberystwyth, Rhayader and Llanilar sheets to the south. The effects of the Devensian ice sheet, during the last glacial period, are reflected in the land forms and the drift deposits. Within the district there are numerous sources for hard rock aggregate although, because of the National Park, there is little likelihood that they will be further exploited.

I am confident that this account will be of interest not only to the professional geologist but also to the increasing number of people who are keen to understand the processes involved in the development our landscape. Finally I must acknowledge that the work could not have been accomplished without the invaluable co-operation of local people who allowed access for my staff to survey their land. I trust that they will receive some satisfaction from this account of the geological origin of this magnificent part of Wales.

Peter J Cook, DSc
Director

British Geological Survey,
Kingsley Dunham Centre,
Keyworth,
Nottingham
NG12 5GG

January 1995

HISTORY OF SURVEY OF THE CADAIR IDRIS SHEET

The primary survey of the district at the scale of one inch to one mile was accomplished by A C Ramsay and A R C Selwyn and published on Old Series Sheets 59SE in 1848 and 59NE in 1850. The descriptions of the geology by these surveyors were incorporated into Ramsay's memoir on the Geology of North Wales in the 1st edition, 1866, and the 2nd edition, 1881.

The present survey was carried out between 1972 and 1991 and was partly accomplished in collaboration with university researchers. The southern strip of the district, south of gridline 00, was mapped at 1:10 560 scale by Dr R Cave between 1972 and 1988. The Upper Ordovician and Silurian sedimentary sequence was mapped, 1986 to 1989, by Drs M J Leng, W T Pratt and M S Scott from the University of Wales, Aberystwyth, under a NERC/University Mapping contract supervised by Drs W R Fitches and R Cave. The Cambrian and Ordovician sequence in the north of the district was mapped, 1988 to 1991, by Drs R A B Bazley, M F Howells, W T Pratt and D G Woodhall. The eastern strip, east of easting 80, was surveyed by Dr W T Pratt.

1:10 000 scale geological maps included wholly or partly in Geological Sheet 149 (Cadair Idris) are listed below with the initials of the surveyors and the dates of survey. Dyeline copies of these maps are available for purchase or public reference in the libraries of the British Geological Survey in Keyworth and Edinburgh. Maps with accompanying Open File Reports are indicated with an asterisk. These are also available for purchase

SH 50 NE	RABB, WTP	1986–87
SH 50 SE *	WTP, RABB	1986–87
SH 60 NW *	WTP, MFH	1987 and 1991
SH 60 NE *	WTP, MJL	1986–67 and 1991
SH 60 SE *	MSS, MJL	1986–67
SH 60 SW *	MJL, WTP	1986–67
SH 61 NW	DGW, WTP	1989 and 1991
SH 61 NE	DGW	1989
SH 61 SE	MFH, WTP	1988–89
SH 61 SW	WTP	1991
SH 70 NW *	MJL, WTP, MSS	1987
SH 70 NE *	MJL, MSS	1987
SH 70 SE *	MSS, RC	1987–88
SH 70 SW *	MSS, RC	1984–87
SH 71 NW	DGW, MFH	1990–1
SH 71 NE	DGW, WTP	1988 and 1990
SH 71 SE *	WTP, MJL, MSS	1987–88
SH 71 SW *	WTP, DGW, MFH	1988 and 1991
SH 80 SW	WTP	1991
SH 80 NW	WTP	1991
SH 81 SW	WTP	1991
SH 81 NW	DGW, WTP	1989 and 1991
SN 59 NE	RC	1973, 1977 and 1988
SN 69 NW	RC	1972–74
SN 69 NE	RC	1972–77 and 1988
SN 79 NW	RC	1966–72
SN 89 NW	RC, WTP	1972 and 1991

ACKNOWLEDGEMENTS

The memoir has been written mostly by Drs W T Pratt, D G Woodhall and M F Howells, but overall it has been very much a collaborative effort, with palaeontological contributions by Drs A W A Rushton, S G Molyneux, J A Zalasiewicz and Mr S P Tunnicliff. Earlier contributions by Drs M J Leng and M S Scott, on the Ashgill and Silurian strata respectively, have also been incorporated. Dr D K Loydell, of University of Wales, Aberystwyth, provided preliminary identifications of graptolites collected during the University contract. Dr R Cave provided guidance on all aspects of the broader regional geology. The geophysical contribution was by Dr R M Carruthers. The memoir was compiled by Dr M F Howells and edited by Drs R W Gallois and A A Jackson.

NOTES

In this book, the word 'district' means the area depicted on 1:50 000 Geological Sheet 149 (Cadair Idris) but excludes the Barmouth peninsula, which is included in Sheet 135 (Harlech) (Figure 1).

Numbers in square brackets are National Grid references. Unless otherwise indicated, all grid references lie within the 100 km square SH.

Numbers preceded by the letter A refer to the British Geological Survey Photograph Collection (other photographs were taken by field staff or university researchers). Numbers preceded by the letter E refer to the Sliced Rock Collection of the Survey. Thin sections without E numbers are in the collection of the University of Wales, Aberystwyth.

Enquiries concerning geological data for the district should be addressed to the British Geological Survey, Keyworth, Nottingham, NG12 5GG.

Welsh words

The following list gives the translations of the most common Welsh words used in this memoir:

afon	river
bryn	hill
bwlch	pass
coed	wood
craig/graig	rock
cwm	valley or cirque
esgair	ridge
llyn/llynau	lake/lakes
nant	stream
pant	small valley or hollow
waun/waen	bog

ONE

Introduction

The district refers to the area between the Dyfi and Mawddach estuaries and includes the southern part of the Snowdonia National Park (Figures 1 and 2). The area of the Barmouth peninsula which encroaches the northern edge of the map was incorporated into the Harlech, 1:50 000 Geological Sheet 135, and described in the accompanying memoir (Allen and Jackson, 1985). Inland of the coastal strip between Tywyn and Arthog, the district is mountainous and sparsely populated. The main towns are Machynlleth, Tywyn and, at the northern edge of the district, part of Dolgellau.

The dominant physical feature is the north-facing escarpment of Cadair Idris with the peaks of Mynydd Moel (863 m above OD), Penygadair (893 m above OD) and Tyrrau Mawr (661 m above OD) (Figure 2). In most places the ground rises sharply from the coast and estuaries, but around Tywyn the rise is more gradual. The most rugged topography is associated with the outcrop of the Ordovician volcanic rocks and is particularly well displayed in the ice-sculpted, serrated profiles around Llynnau Cregennen and Kings. The extremely prominent north-east–south-west valley between Tal-y-llyn Pass and Bryncrug to the south-west, which is partly occupied by the Afon Dysynni, lies along the Tal-y-llyn Fault, a branch of the Bala Fault. To the south, the Silurian strata typically form smooth rounded hills although in the south-east of the district there is a fairly prominent and persistent plateau at about 250 to 300 m above OD.

To the north of the Cadair Idris escarpment, the main drainage is into the Mawddach estuary, at the northern edge of the district (Figure 2). In the south, the district is drained by the Afon Dysynni and Afon Dyfi and their complex of tributaries; the Dyfi estuary lies south-west of Machynlleth, to the south of the district.

Hill farming, forestry and tourism are the main industries. In the past, in villages such as Corris, Aberllefenni and Abergynolwyn, slate production was a major feature of the local economy. However, the small quarries in these villages were some of the first to fail during the slump in the slate industry during the early years of this century. All attempts to create a renaissance have failed and the only current activity is small scale, at Aberllefenni, for architectural slabs.

PREVIOUS RESEARCH

Systematic and considered observations on the geology of north-west Wales began with those of Sedgwick (1844, 1852) and Sedgwick and Murchison (1835). Sedgwick published his map of the whole of North Wales in 1845 and this was followed by a larger-scale map, produced by

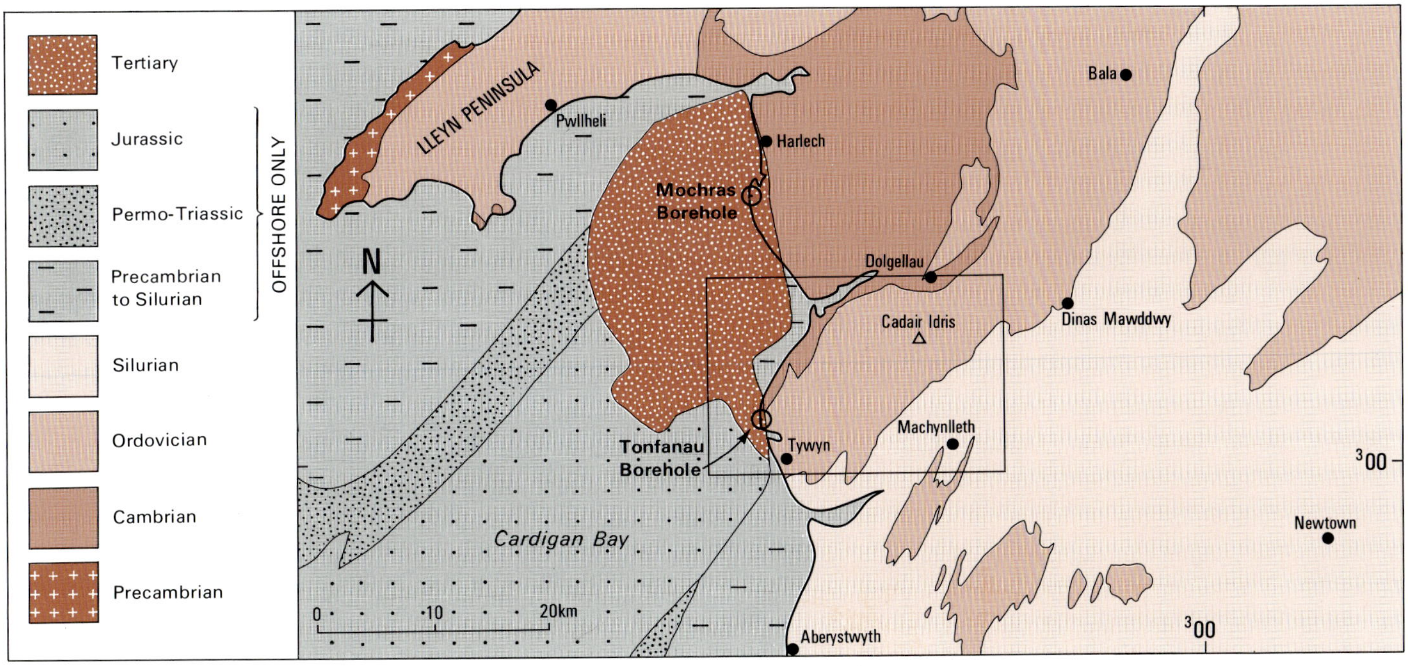

Figure 1 Simplified geology of the Cadair Idris Sheet (149) and surrounding area.

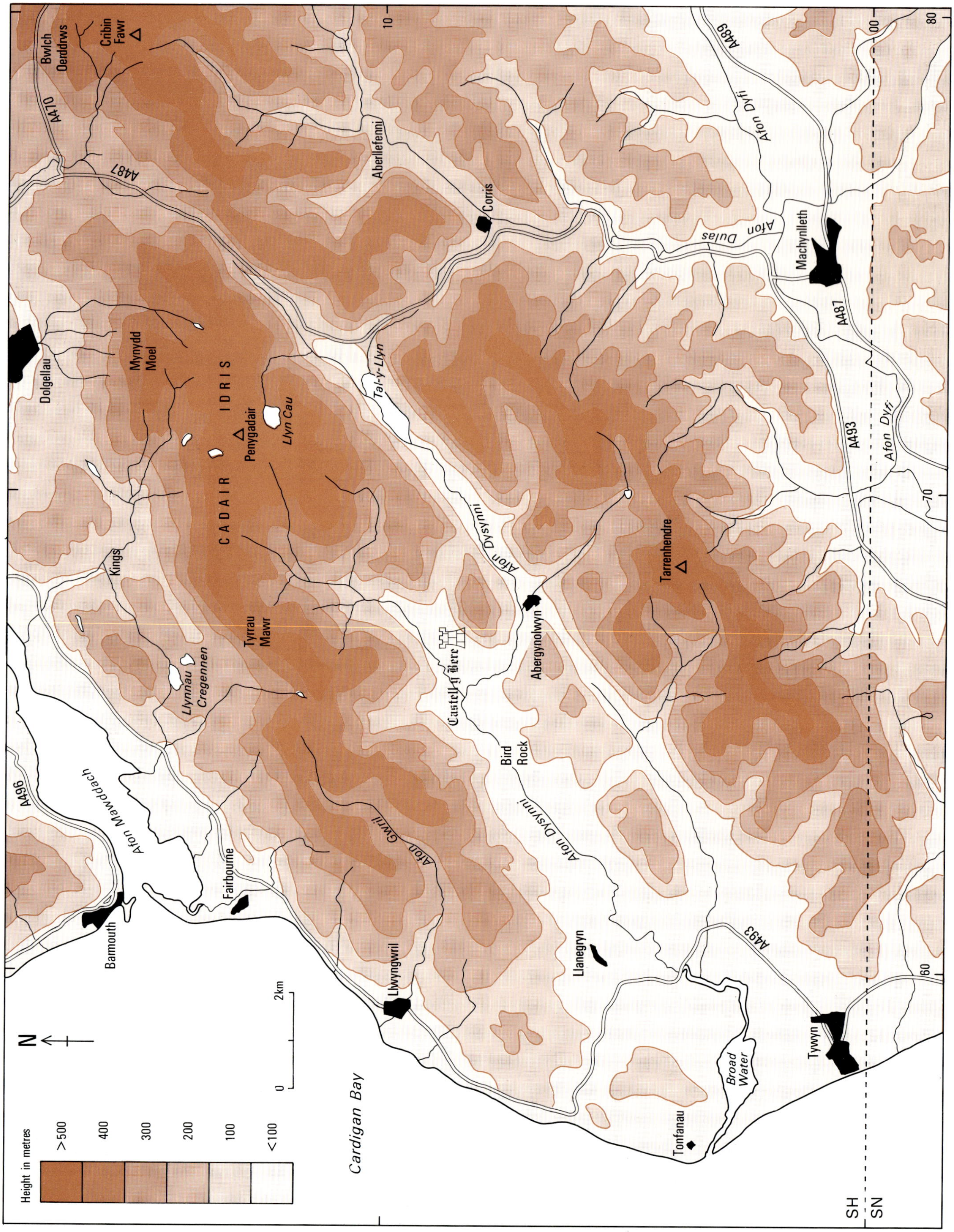

Figure 2 Physical features of the Cadair Idris district.

Daniel Sharpe, in 1846, the year that the Geological Survey established their presence in the area. The Geological Survey maps (Old Series one-inch Sheets 59NE/SE) of the area around Cadair Idris were published in 1848 and 1850 and many descriptions of the geology of the district were incorporated into Ramsay's memoir (1866, 1891) on the geology of North Wales. All of this work was accomplished prior to Lapworth's (1879) establishment of the Ordovician System, which replaced Sedgwick's 'Upper Cambrian' and Murchison's 'Lower Silurian'. Within the district, the clear determination of the stratigraphical sequence was not completed until the work of Matley and Wilson (1946) on the Cambrian sequence of the Harlech Dome; Cox (1925), Cox and Wells (1921, 1927) on the Cambrian–Ordovician sequence of the Cadair Idris escarpment, and Jones and Pugh (1916) on the Silurian strata. These works include descriptions of the extrusive and intrusive volcanic rocks in the sequence, as does the work of Wells (1925) and Davies (1959) but, until this current survey, there has been little interpretation in the light of current understanding of volcanic and sedimentary processes.

GEOLOGICAL HISTORY

The coarse sandstones of the Cambrian, Harlech Grits Group are well exposed on the Barmouth peninsula but concealed beneath drift on the Fairbourne spit, south of the Mawddach estuary (Figures 2 and 3). They represent a period of major, proximal, coarse-grained, turbiditic sedimentation in this part of the Welsh Basin (Allen and Jackson, 1985). The Gamlan Formation, at the top of the group, is the oldest formation exposed south of the Mawddach estuary and comprises a sequence of grey and purple silty mudstones and volcaniclastic rocks with thin manganese-enriched laminae.

The Clogau Formation, at the base of the Mawddach Group, comprises black mudstones, reflecting anaerobic sea-floor conditions, and a few massive bedded turbidite sandstones. The formation represents a key episode in basin development as all subsequent deposition was finer grained. Such a change is most clearly expressed in the distal, fine-grained sandstone and siltstone turbidites of the Maentwrog Formation which probably marked the end of the basinal stage (Allen and Jackson, 1985).

The Ffestiniog Flags Formation, of pale grey silty mudstone with flaggy quartzose siltstone beds, is considered to have accumulated on an open shelf, locally above storm-wave base. The overlying carbonaceous black mudstones of the Dolgellau Formation mark a temporary return to poorly oxygenated bottom waters, prior to the establishment of more open-shelf conditions in Tremadoc times, when silty, locally bioturbated muds (Dol-cyn-afon Formation) were deposited. In spite of the proximity to Rhobell Fawr, a site of major Tremadoc volcanism, the Dol-cyn-afon Formation bears little evidence of volcaniclastic contamination and the Rhobell Volcanic Group is thin and restricted to Dolgellau (Figure 3).

The Aran Volcanic Group, which overlies the Mawddach and Rhobell Volcanic groups, records an episode of volcanism and sedimentation which lasted from Arenig to Caradoc times. The basal, Allt Lŵyd Formation consists of shallow-water sandstones and conglomerates which are generally conformable upon the Dol-cyn-afon Formation. Towards the northern edge of the district, closer to Rhobell Fawr, there is clear evidence of overstep of the sandstones on to the Ffestiniog Flags Formation, indicating pre-Arenig uplift and erosion. The sandstones and conglomerates indicate a possible derivation from both the uplifted Harlech Grits Group, and the Rhobell Volcanic Group, to the north of the district.

Above the Allt Lŵyd Formation, the component formations of the Aran Volcanic Group indicate a complex cycle of acid and basic volcanic activity mainly associated with outer shelf, mudstone–siltstone deposition with much incorporation of volcaniclastic debris. A temporary shallowing, to within wave base, caused deposition of an oolitic ironstone near the middle of the sequence. The volcanic activity was dominated by the eruption of basalt lavas with much evidence, such as pillows, quenched and spalled surfaces, and wet-sediment/lava interaction, of subaqueous emplacement. Throughout the sequence there is evidence of acid volcanism, in primary acidic ash-flow tuffs and in debris flow deposits. The last phase of volcanism, represented by the Craig Cau Formation, was dominated by major acidic ash-flow tuff eruption. Within the district, there is much evidence of post-emplacement slumping that was possibly related to late-stage rhyolite intrusion and the disruption of an unlithified substrate.

The top of the volcanic sequence shows little evidence of reworking and the overlying Caradoc sedimentary rocks are mainly turbiditic, silty mudstones with evidence of some slumping. This pattern was broken at the end of Caradoc times with the deposition of the black, pyritous muds of the Nod Glas Formation which represent an extensive, anaerobic basinal event in Wales. Significant shallowing occurred in the late Ashgill, and influxes into the district of coarse clastic debris, the Garnedd-wen Formation, probably reflect a worldwide, low sea-level stand. The Llandovery deposits are mainly fine-grained sandstone and mudstone turbidites, with intercalated hemipelagic mudstones. The slight lithological differences between the Llandovery formations indicate the delicate balance of these sedimentation processes and, in the uppermost formations, this is compounded by their marked diachroneity.

The main structural overprint, and the low grade metamorphism of the rocks within the district, is almost entirely due to the end-Silurian–early-Devonian, end-Caledonian deformation. The metamorphic grade implies a considerable depth of burial and the subsequent removal by erosion of a few kilometres of Silurian and Devonian strata. Some of the north–south orientated folds, mainly in the Cambrian–Ordovician sequence, may be inherited from an earlier period of compression and uplift in the area of the Harlech Dome, to the north of the district. Within the district there is little evidence to suggest that either the Bala or Tal-y-llyn faults were active during Lower Palaeozoic times.

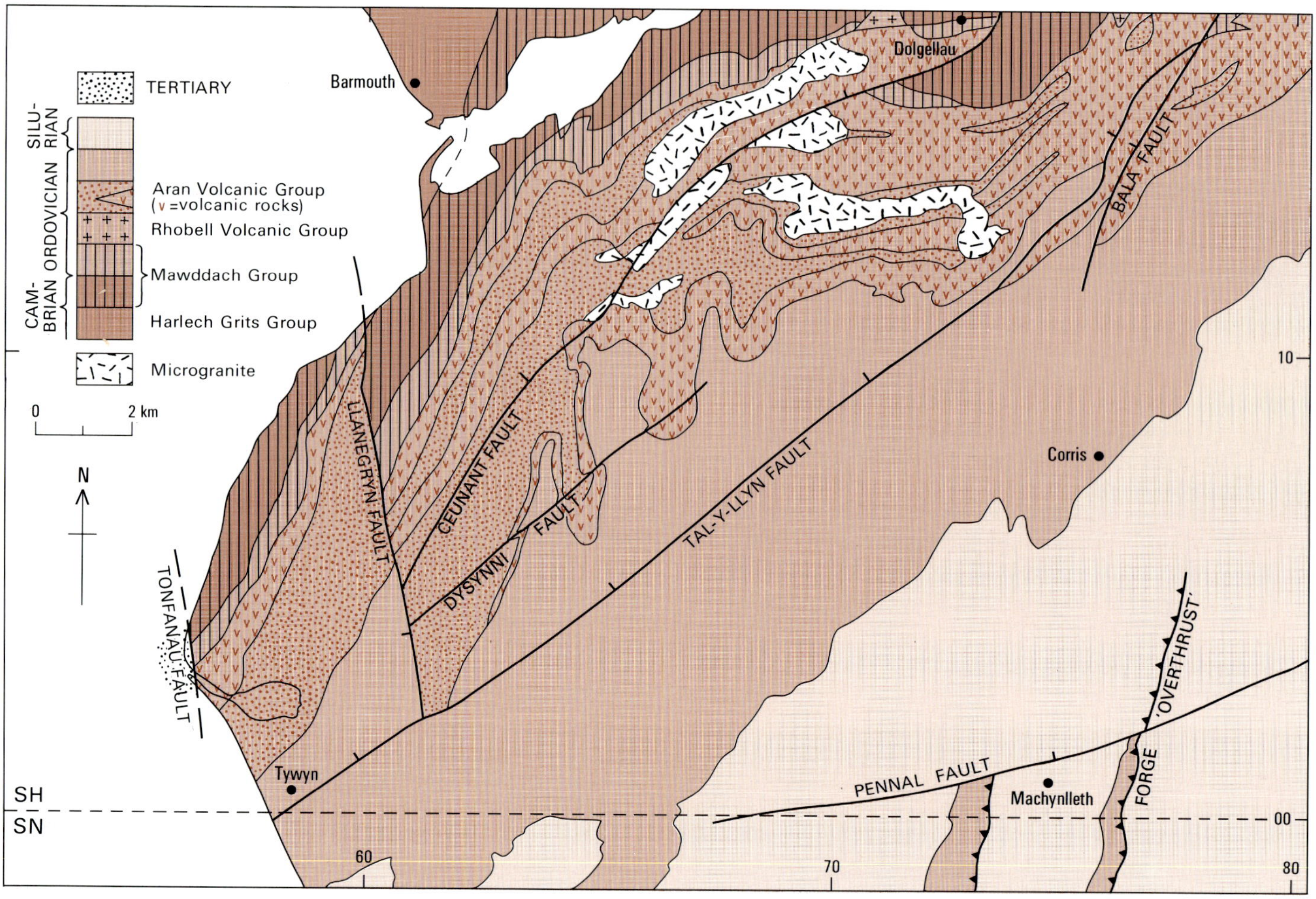

Figure 3 Geological sketch map of the Cadair Idris district.

Upper Palaeozoic, Mesozoic and Tertiary rocks are nowhere exposed in the district, but, by comparison with dated sequences in the Mochras Borehole to the north, strata in a borehole at Tonfanau have been ascribed a Tertiary age. Mesozoic strata, of Middle Jurassic age, may also be present beneath the Quaternary in the coastal strip since they crop out a short distance offshore.

The Quaternary (Devensian) deposits are entirely the products of the most recent glaciation and occur throughout the district. The erosional effects of the glaciation are dramatically displayed in the high cwms and in the deeply scoured valleys. It is almost certain that at some time the district was entirely covered by ice. The deposits are mainly of till, comprised mostly of local debris. However, in the coastal tract, till deposited by the Irish Sea ice contains clasts of Mesozoic age. In postglacial times, the silting-up of the major estuaries was facilitated by the growth of spits and, more recently, this process has been accelerated by land reclamation.

TWO

Cambrian

The Middle and Upper Cambrian strata, which crop out in the north-west and north of the district, lie at the southern edge of the classic Cambrian sequence of the Harlech Dome (Belt, 1867–68; Andrew, 1910; Matley and Wilson, 1946). The strata north of the Mawddach estuary, around Barmouth, were incorporated into the geological map of the Harlech district (Institute of Geological Sciences, 1982) and described by Allen and Jackson (1985). South of the estuary, the Cambrian strata crop out mainly in a coastal strip between Tonfanau and Barmouth (Figures 2 and 3). Apart from in the cliffs between Llwyngwril and Fairbourne, exposure is generally poor. The sequence, which comprises mainly turbiditic siltstones and mudstones with black laminated, hemipelagic mudstones (Figure 4), records steady subsidence and marine sedimentation, in well-oxygenated, shallow-marine conditions alternating with periods of deeper, poorly oxygenated conditions below storm wavebase.

HARLECH GRITS GROUP

This group comprises the strata between the Bryn-teg Volcanic Formation (Allen and Jackson, 1978), of late Precambrian or Early Cambrian age, and the Clogau Formation. The group is dominated by coarse sandstone, but also includes silty mudstones and manganiferous beds. Only the Rhinog, Hafotty, Barmouth and Gamlan formations are exposed within the district, and only the Gamlan Formation is exposed south of the Mawddach estuary (Figures 3 and 4).

Gamlan Formation

The Gamlan Formation, broadly the equivalent of the Solva Group of Pembrokeshire (Harkness and Hicks, 1871; Hicks, 1881; Cowie et al., 1972), crops out in a small area east of the Llanegryn Fault and is exposed only in the coastal section at Llwyngwril (Figures 4, 5). In the Harlech district the formation is 200–300 m thick (Allen and Jackson, 1985), but at Llwyngwril only the uppermost 30–40 m are exposed. No fossils were recovered during the survey.

The lowest exposed strata at Llwyngwril [5992 1105] comprise pale grey and green silty mudstones, 0.5–0.2 m thick, with widely spaced, weakly graded siltstone beds, less than 10 mm thick. Bedding in the mudstones is poorly defined and commonly disrupted by bioturbation, but the siltstone beds display ripple cross-stratification, planar lamination and convolute bedding. Large pyrite cubes, up to 10 mm across, occur throughout the mudstones and small green 'reduction' spots are common. Upwards, the strata become more thinly bedded and siltstone beds more common, and locally disaggregated through bioturbation. The mudstones are striped pale grey and purple, and the siltstones are pale green (Plate 1a). However, the colour boundaries, which are defined sharply by haematite-rich bands less than 1 mm thick, commonly cut across lithological boundaries, indicating their secondary, diagenetic origin.

The upper part of the formation contains many small diagenetic concretions, generally less than 10 mm in diameter, comprising a shell of very fine-grained quartz and manganese-garnet about a core of coarse-grained chlorite and quartz (see Woodland in Matley and Wilson, 1946, p. 33) (Plate 1a). The concretions display a great variety of shapes, including ellipsoidal, U-shaped and tubular types, and their association with disrupted beds suggest that they probably nucleated on worm burrows (cf. Matley and Wilson, 1946), although others (Price, 1963) have preferred an inorganic origin. All the concretions are deformed and deflect the regional cleavage. In places, the concretions are amalgamated into sheets, up to 3 mm thick, at the base of some siltstone beds (Plate 1b).

The uppermost 10 m of the formation are thinly bedded, purple and grey silty mudstones and green laminated siltstones. Widely spaced, massive, silty mudstone beds, up to 0.5 m thick, are crowded with concretions and manganese staining is particularly prominent on weathered surfaces, along cleavage and around concretions. Some thin, fine-grained siliceous beds contain concentrations of manganese garnet ('coticules' of Renard, 1878). A prominent layer, with cone-in-cone concretions up to 0.6 m long and 0.1 m thick, occurs in the top metre of the formation (Figures 5 and 6).

Conditions of deposition

The strata are interpreted as the deposits of low density, mud-dominated turbidity currents. The absence of shallow-water sedimentary structures implies that they were laid down beneath wave base, but Crimes (1970) noted that rich trilobite faunas from the upper part of the equivalent strata on St Tudwal's peninsula showed little evidence of transport, and, therefore, no great depth. Bioturbation implies oxygenated bottom waters, but the strong colour contrasts suggest variable early diagenesis, fluctuating between oxidising conditions, which fixed haematite and resulted in a purple colouration, and reducing conditions, which fixed pyrite and produced a green (chlorite) colouration.

The distinctive diagenetic manganiferous minerals are the same as those of the Manganese ore bed of the older Hafotty Formation, around Barmouth (Woodland, 1939; Price, 1963; Mohr, 1964). Allen and Jackson (1985) and Bennett (1987) have described relict volcanic textures from the coticules in the Gamlan Formation, suggesting

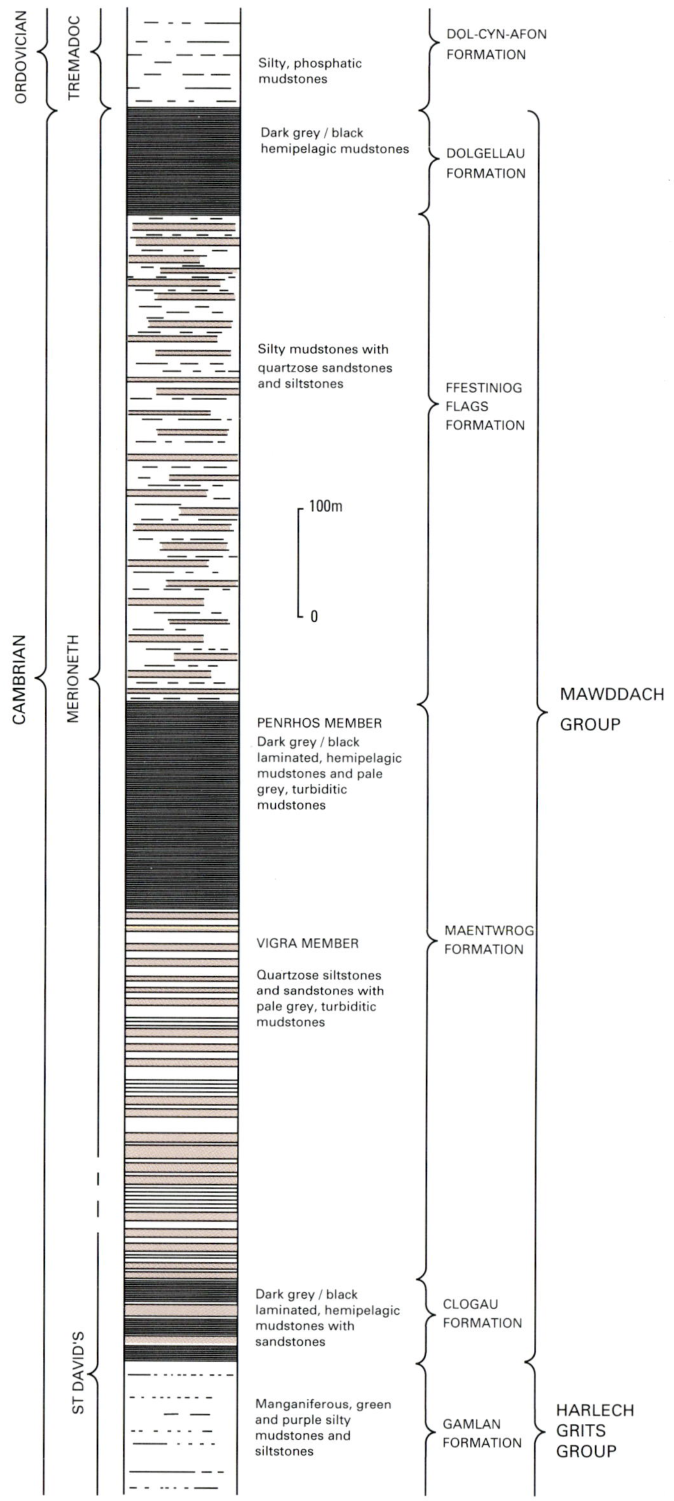

Figure 4 Simplified stratigraphy of the Cambrian sequence.

a similar volcanic, possibly exhalative, origin for the manganese concentration. However, the proportion of manganiferous minerals is much lower in the Gamlan Formation, well below potential ore grade, probably because of greater dilution by incorporated sediment.

MAWDDACH GROUP

The Mawddach Group represents the second phase of Cambrian sedimentation and is dominated by fine-grained strata, mainly siltstones and silty mudstones, with few coarse-grained sandstones. Laminated hemipelagic mudstones are the oldest in the Lower Palaeozoic sequence. The stratigraphy broadly follows Allen et al. (1981), but the Cwmhesgen Formation is abandoned and its former components, the Dolgellau (Upper Cambrian) and Dol-cyn-afon (Tremadoc) members, are defined here as formations (Figure 4). The biostratigraphy of the group is similar to that of the Harlech District (Allen et al., 1981), but is less well known.

Clogau Formation

This formation comprises about 80 m of turbiditic mudstone, siltstone and sandstone, with interbedded black, laminated, hemipelagic mudstone, rich in framboidal pyrite. It is broadly equivalent to the Menevian of Belt (1867–68), the Menevian Beds of Jones (1933) and the Clogau Shales of Andrew (1910). It crops out only in a small area near Llwyngwril, but this coastal section (Figure 5; Jones, 1933) is the best exposure of the formation around the Harlech Dome. The proportion of sandstone and silty mudstone at Llwyngwril is much greater than in the equivalent sequence at Barmouth and on the east side of the Harlech Dome (Allen et al., 1981; Allen and Jackson, 1985), and black mudstones form less than 20 per cent of the total thickness.

The formation can be divided into upper and lower divisions. The lower division, 52 m thick, comprises varying proportions of thinly bedded black mudstone and turbiditic silty mudstone, with three discrete sandstone-rich sequences (Figure 5). The lowest of the three sandstones occurs at a very similar stratigraphical position to the Cefn Coch Grit of the Harlech Dome (Andrew, 1910) and may be its lateral equivalent. The upper division of the formation, 25 m thick, is rich in black mudstone, with few sandstones and is similar to the lithologies in the Harlech district. The upper part is exposed in road cuttings [6000 1099] and in the course of the Afon Caletwr [6001 1102].

The base of the formation is well exposed in the coastal cliff near the mouth of the Afon Caletwr [6001 1116] and can be traced for approximately 200 m to the north-east [6015 1129]. Between these localities, the succession is broadly similar, with the Gamlan Formation overlain by 1.5–2 m of transitional beds which comprise alternations of silty mudstone with manganiferous nodules, typical of the Gamlan Formation, and black, laminated mudstone with interbeds of pale grey turbiditic mudstone and siltstone, typical of the Clogau Formation. A layer of cone-in-cone concretions, each up to 0.5 m long and 0.08 m thick, locally occurs 0.55 m above the base. The basal beds are overlain by about 4 m of thinly bedded black mudstone and silty mudstone with impersistent sandstone beds up to 0.6 m thick.

A sequence of thickly to thinly bedded, turbiditic quartzose sandstone, some 10 m thick, which begins 6 m above

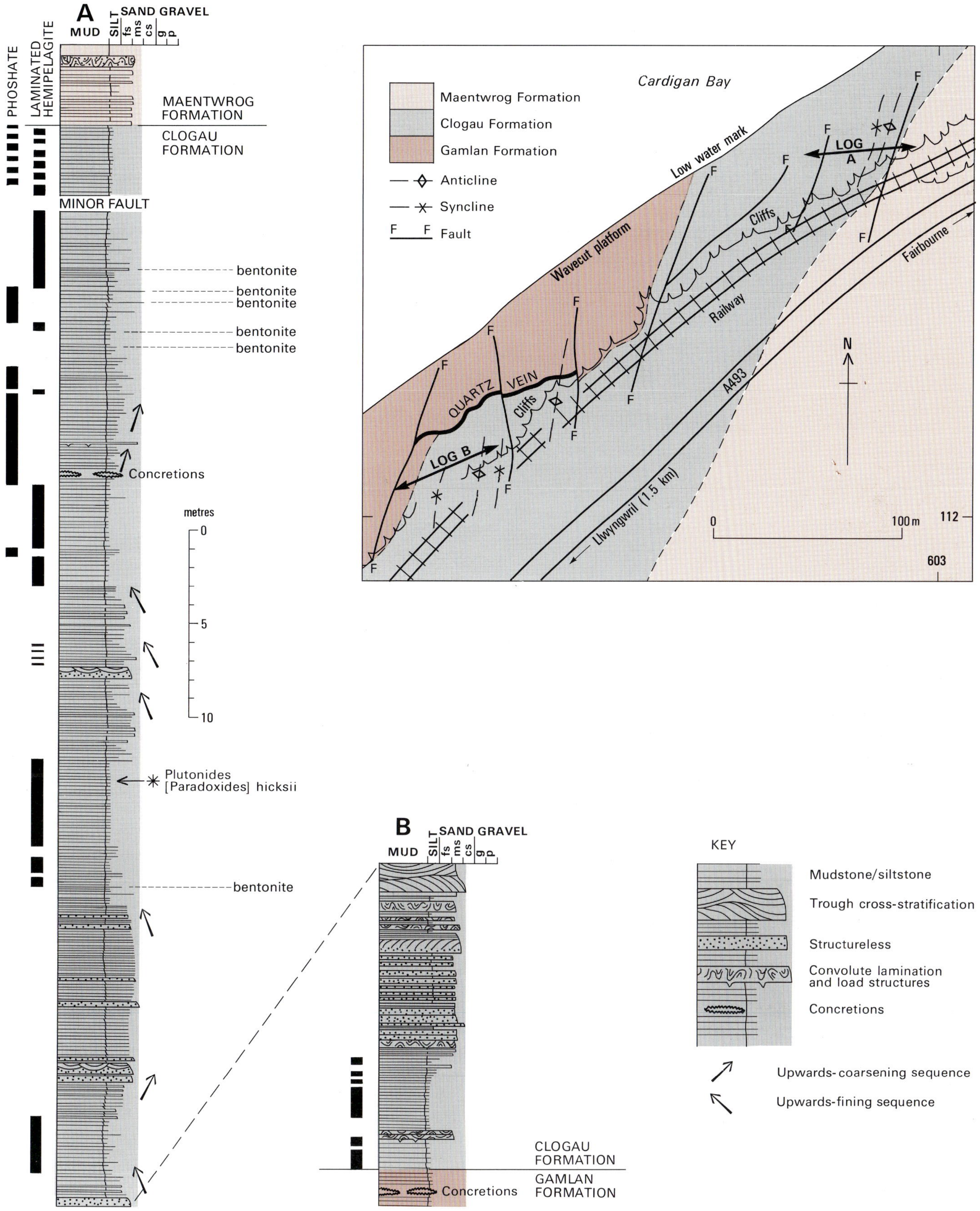

Figure 5 Detailed sedimentary logs through the Clogau Formation at Llwyngwril [6007 1122]. Inset map shows location of logged strata.

Plate 1a Horizontal surface displaying purple and green silty mudstones with quartz/manganese garnet concretions. Key for scale. Gamlan Formation. Llwyngwril [6000 1113].

Plate 1b Horizontal surface displaying mudstones and silty mudstones with concretions, minor faults and tightly folded, pink weathered concretionary layers (top of picture). Gamlan Formation. Llwyngwril [6000 1113].

the base of the formation, forms an almost continuous exposure, in a series of folds, for a distance of almost 200 m along the coast (Figure 6; Plate 15a). Individual beds display considerable lateral variations in thickness. The thicker sandstone beds have sharp bases, commonly eroded into the underlying strata, and graded basal layers. A great variety of internal bedforms can be distinguished. In some beds the thin, graded basal portion is overlain by trough cross-stratified, medium- to fine-grained sandstone. Other beds comprise structureless sandstone overlain by convoluted bedding and, in turn, by large-scale cross-stratification. The thickest sandstone beds, up to 2 m, are an amalgamation of several sandstone turbidites which pass laterally into turbiditic sandstones with interbedded mudstones. Within some composite sandstone beds, steep-sided channels, some with undercut margins, are infilled by either cross-stratified sandstone or muddy, high-matrix sandstone with contorted rip-up clasts. Planar beds of high-matrix sandstone, up to 0.7 m thick, with numerous rip-up clasts are a small, but distinctive, element of the sandstone sequence.

The remainder of the lower division of the formation comprises pale grey mudstone, siltstone and black mudstone, with two further sandstone sequences (Figure 5). The individual sandstone beds are much thinner, 20–50 mm, and display less channelling and syndepositional disruption, than those below.

The upper division of the formation comprises interbedded black, laminated mudstone and turbiditic silty mudstone (Plate 1c). Individual beds rarely exceed 30 mm and

Plate 1c Interbedded dark grey/black hemipelagic mudstone and pale grey turbiditic silty mudstone. Clogau Formation. Llwyngwril coast section [6035 1143].

the thickest single bed of black mudstone, 14.5 m below the top of the formation, is only 80 mm thick. Disseminated pyrite and euhedral pyrite crystals occur throughout and are particularly concentrated at bed boundaries. Small, syndepositional, extensional faults are common. Five soft, green talcose beds, up to 0.11 m thick, probably metabentonites, occur close to the top of the formation. In parts, sequences up to 7 m thick are almost entirely of siltstone beds, up to 60 mm thick, and turbiditic mudstone, up to 80 mm thick, and are rich in phosphate nodules, small carbonate concretions and thin concretionary layers.

BIOSTRATIGRAPHY

Stratigraphically significant fossils were found only at one locality [6028 1138], about 36 m below the top of the form-ation. The occurrence of *Plutonides [Paradoxides] hicksii* (together with sponge spicules) (Plate 2, i, j) indicates the presence of the *Tomagnostus fissus* Biozone, mid-St David's Series. The same zone was proved in the lower part of the formation in the east of the Harlech District (Allen et al., 1981, p. 303).

CONDITIONS OF DEPOSITION

The occurrence of laminated hemipelagites and framboidal pyrite implies mainly poorly oxygenated bottom water conditions. However, the phosphatic mudstones in the upper part of the formation indicate that the balance between well, and poorly, oxygenated conditions was easily changed, probably by emplacement of turbidites and accompanying plumes of oxygenated water. The lowest sandstone sequence reflects powerful, high-density turbidity currents and dilute debris flows. The upwards-coarsening and upwards-thickening cycle implies that the sand body represents the infill of a channel. It is unlikely that switching of a channel on an existing submarine fan was responsible. Instead, the sudden influx of sand into 'restricted basin' type sediments probably reflects tectonic or eustatic activity. The two higher sandstone packets, which lack channels and thick sandstones, represent more distal depositional sites and are more typical of a smooth turbidite fan or lobe.

Maentwrog Formation

The Maentwrog Formation, comprising the Vigra Flags and Penrhos Shales of Matley and Wilson (1946), is characterised by white weathered, highly indurated, quartzose and micaceous siltstone beds, the 'ringers' of Fearnsides (1910). It conformably overlies the Clogau Formation and its base is placed at the top of the highest black mudstone and the appearance of quartzose siltstones and fine-grained sandstones, 3–100 mm thick (Figure 5). The formation, 540 m thick, is totally exposed in the Llwyngwril/Fairbourne coastal section, which is defined here as the stratotype, and also exposed in isolated knolls amidst the drift cover of the Mawddach estuary north-east of Fairbourne. The base of the formation is well exposed, east of a quartz veined fault, on the foreshore [6032 1139] between Llwyngwril and Friog. No fossils were recovered during this survey, but in the Harlech District, the *Olenus* Biozone of the Merioneth Series is known at several localities.

VIGRA MEMBER

This member comprises 340 m of thinly bedded, turbiditic pale grey mudstone, quartzose siltstone, fine-grained sandstone and black, laminated, hemipelagic mudstone. The relative proportion and thickness of these elements is variable. Turbiditic mudstones, locally with very thin, planar laminated siltstone bases, are the dominant lithology, but the frequency of black hemipelagic mudstone beds generally increases upwards. The siltstones and fine-grained sandstones, 5–100 mm thick, are planar laminated and/or ripple cross-stratified. Convolute lamination and starved ripples are also common.

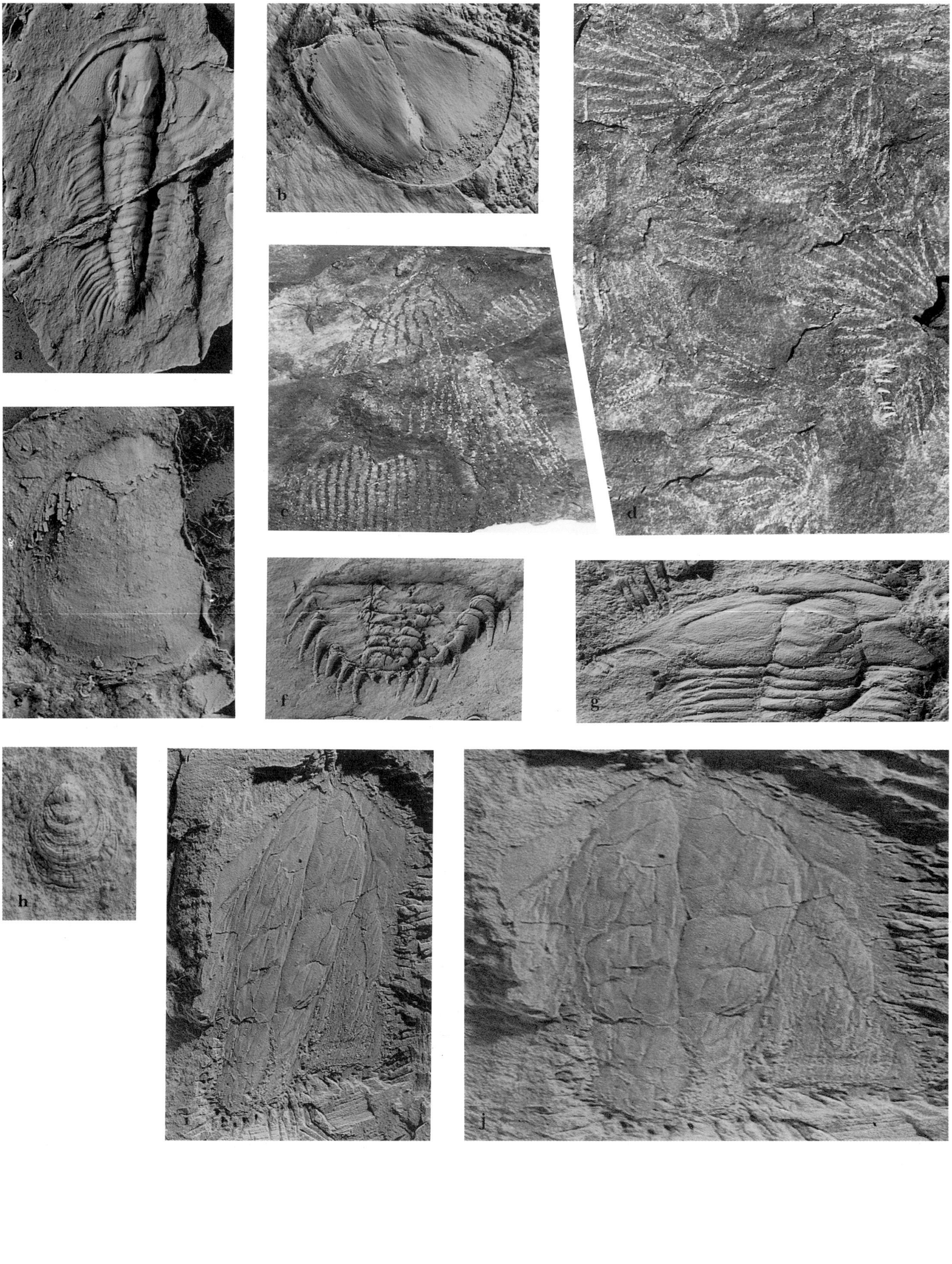
a
b
c
d
e
f
g
h
i
j

Plate 2 Fossils from the Mawddach Group.

a) *Pseudokainella impar* (Salter), latex cast, RU 7071, × 2. Upper Dol-cyn-afon Formation (upper Tremadoc, *pusilla* Biozone), Llyn Wylfa [6699 1630]. b) *Asaphellus homfrayi* (Salter), pygidium, RX 3209, × 1.5. Upper Dol-cyn-afon Formation, Llyn Wylfa [6697 1630]. c) *Rhabdinopora flabelliformis flabelliformis* (Eichwald, sensu Bulman), RX 3197, × 1.5. From the "lower *Dictyonema* Band" of Cox & Wells (1921). Lower Dol-cyn-afon Formation (lower Tremadoc, *flabelliformis* Biozone), Nant Gwynant [6805 1674]. d) *Rhabdinopora flabelliformis anglica* (Bulman), RU 7051, × 1.5. From the "upper *Dictyonema* Band". Upper Dol-cyn-afon Formation (upper Tremadoc, *pusilla* Biozone?), Nant Gwynant [6828 1654]. e) *Lingulella davisii* (McCoy), latex cast of intaglio of brachial valve (mainly an internal mould), RU 6941, × 3. Ffestiniog Flags Formation (Merioneth Series), Gwern-y-barcud [7001 1766]. f) *Parabolina spinulosa* (Wahlenberg), latex cast of pygidium, RU 6961, × 3. Merioneth Series (*spinulosa* Biozone), Dolgellau Formation, Gwern-y-barcud [6998 1754]. g) *Parabolinoides bucephalus* (Belt), cranidium with displaced free cheek and part of thorax, RU 6936, × 3. Merioneth Series (*spinulosa* Biozone?), uppermost Ffestiniog Flags Formation, Gwern-y-barcud [7000 1762]. h) *Lingulella* sp., latex cast of pedicle valve RU 6946, × 3. This form, smaller and more pointed posteriorly than *L. davisii*, characterises the passage beds from Ffestiniog Flags to Dolgellau formation. East of Gwern-y-barcud [7012 1773]. i) j) *Plutonides hicksii* (Salter), RX 3881. i) compressed cranidium, × 1.5. j) the same with the distortion removed (Rushton & Smith, 1993). Clogau Formation (St David's Series, *fissus* Biozone?), Llwyngwril [6024 1137].

Ripple foresets, flutes and groove casts indicate northerly transport of turbidity currents (Crimes, 1970a). About 15–20 m below the top of the member, a dominantly black mudstone sequence, 8 m thick, is well exposed near Friog [6084 1182] and is lithologically similar to the upper part of the Clogau Formation.

PENRHOS MEMBER

The member, which is 200 m thick, is characterised by approximately equal proportions of hemipelagic black mudstone and turbiditic mudstone. The base is marked [6092 1187] by a layer of black mudstone, 0.7 m thick. The mudstones are well cleaved, soft and, because of a high pyrite content, typically ochreous weathered. Sharp-based siltstone turbidites, which grade into thicker, pale grey turbiditic mudstone, occur throughout, but rarely exceed 10 mm and are mainly 1–2 mm thick. Bedding-parallel carbonate concretions are common. The member is best exposed on the coast at Friog [6102 1192] and in a nearby road cutting [6121 1202].

CONDITIONS OF DEPOSITION

The hemipelagic black mudstones of the formation suggest sustained periods of poorly oxygenated conditions and there is little indication of any change in water depths and palaeogeography since those established during deposition of the Clogau Formation. The dominance of quartz within the siltstones and sandstones implies sustained working of material in a shallow-marine environment, prior to incorporation into turbidites, and derivation from a source region of metamorphic and igneous rocks (Woodland, in Matley and Wilson, 1946). Crimes (1970), noting the similarity of rocks of this age throughout Wales, indicated that the sediment source probably lay to the south-west of Wales.

Ffestiniog Flags Formation

The Ffestiniog Flags Formation conformably overlies the Maentwrog Formation and is about 460 m thick near Fairbourne, and at least 500 m thick south of Dolgellau. The formation is typically well exposed and well featured and it forms knolls surrounded by alluvium in the Mawddach estuary [6300 1460 and 6385 1540]. Good exposure also occurs in Friog quarries [6193 1223] and in cuttings along the former Penmaenpool to Arthog railway e.g. [6540 1645]. The base of the formation is not exposed in the district, but exposures within a few metres of it, near Friog [6106 1165], indicate that it is relatively sharp and marked by the appearance of siltstones.

The formation is mainly of pale grey or green, poorly cleaved silty mudstone with streaks and discontinuous ripples of siltstone, some faint lamination and traces of vertical bioturbation, and quartzose siltstone/fine-grained sandstone beds, generally 10–100 mm thick, but up to 0.4 m. In places, sandstones increase in frequency and dominate sequences 10–20 m thick. The sandstone beds have sharp bases, commonly loaded, and generally have well-defined contacts with the overlying silty mudstone. Most beds are planar laminated, locally with ripple cross-stratified tops, and are not noticeably graded, but others display tabular cross-stratification or convolute lamination throughout. Large-scale, hummocky cross-stratification has been noted in some beds. In others, large-scale cross-bedding, convolute lamination or slump folds in the lower part are truncated above by planar laminated siltstone. Ripple crests, both symmetrical and asymmetrical, and locally interfering sets, are well exposed on some bedding planes. Tabular sandstone beds can be traced laterally for tens of metres, but shallow channels, infilled by cross-stratified sandstone, also occur. The sandstones are dominated by quartz grains and contain less than 5 per cent feldspar (E65193, E65195); grain margins are indistinct because of recrystallisation. Detrital muscovite, commonly with thin leaves of chlorite, is a persistent accessory.

Thin sequences of thickly bedded, brown weathered sandstone, 3–7 m thick, occur within the uppermost 50–100 m of the formation. Individual beds are impersistent and the sequences are lenticular, probably channel infills, for example Fegla Fawr [6335 1496]. At Friog quarries [6193 1218], two such sequences, each about 7 m thick, are separated by 13 m of thinly bedded siltstones and silty mudstones. The sandstones are mainly trough cross-stratified, but planar lamination, internal scours and slump folds also occur. Normal grading is only weakly developed and some sandstones amalgamate along strike into thicker, composite beds. On the base of some beds are meandering, horizontal bioturbation traces and tool mark casts, the latter indicating southward-directed palaeocurrents. The sandstones are micaceous and, because of a higher propor-

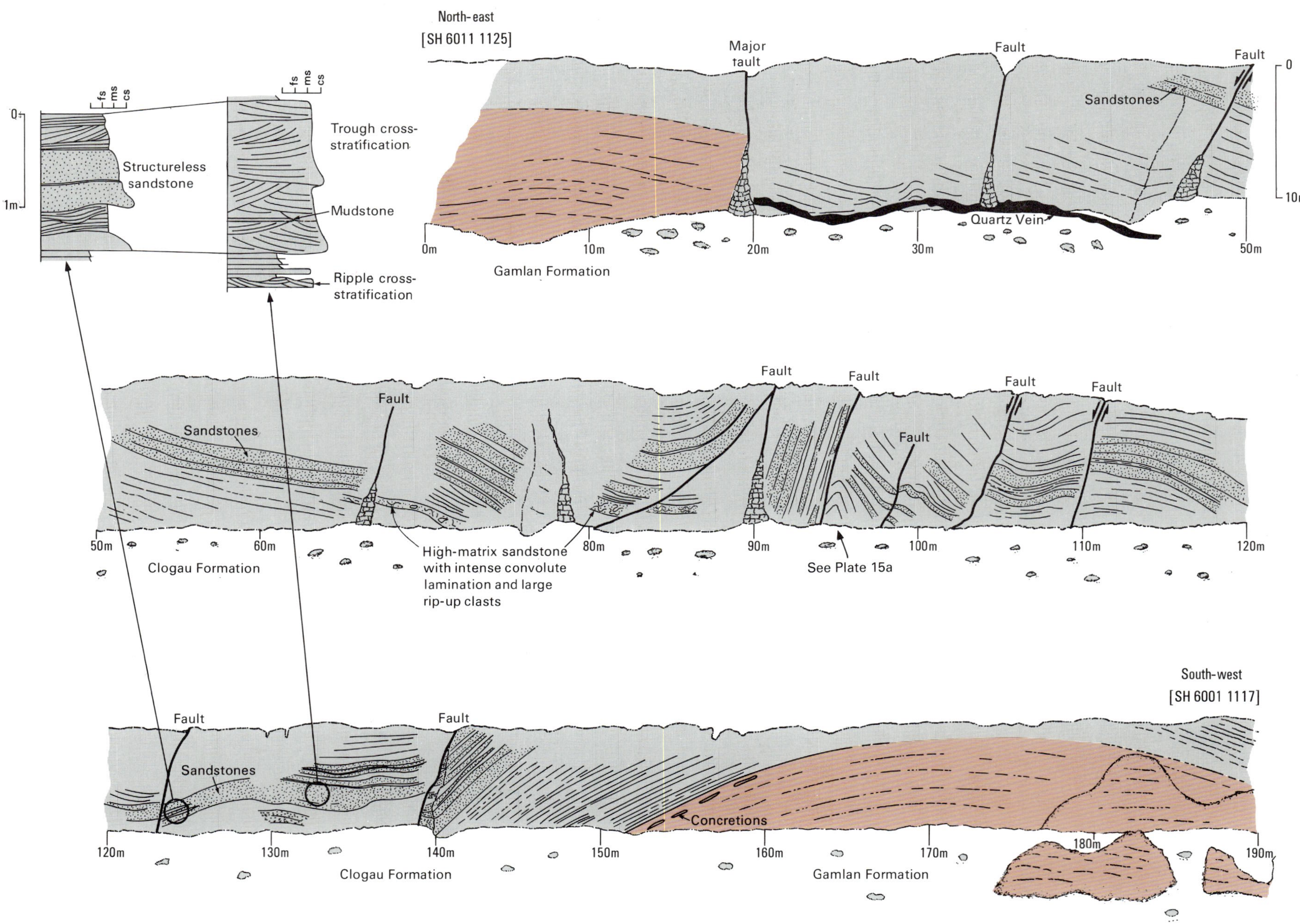

Figure 6 Sketch of the cliffs at the Llwyngwril coastal section [6007 1122]. Small-scale sedimentary logs illustrate the lateral variation in the sandstones near the base of the Clogau Formation.

tion of calcareous and chloritic cement, have a much less indurated appearance than the quartzose siltstones/sandstones of the formation (E62023, E62026).

BIOSTRATIGRAPHY

This formation, particularly the upper part, is characterised by the brachiopod *Lingulella davisii* (Plate 2 e), which occurs either at the bases of thick sandstones, for example at Friog [6181 1210] and Arthog [6526 1595], or within the muddier, transitional strata of the top 50 m, for example at Abergwynant [6782 1716] and Gwern-y-barcud [7001 1766], north-west of Garth Angharad. Although this species is of doubtful stratigraphical value, when present in abundance it is thought to represent shallow, well-oxygenated waters. The transition up into the overlying Dolgellau Formation, seen on the east bank of the Gwynant, resembles that observed in the Harlech District in yielding a smaller form of *Lingulella* (Plate 2 h). The trilobite *Parabolinoides bucephalus* (Plate 2 g) is present at about this level south of Gwern-y-barcud [7000 1762], which is one of the type localities for the species (Belt, 1867–68). This fossil indicates the *Parabolina spinulosa* Biozone of the early Merioneth Series.

CONDITIONS OF DEPOSITION

The silty mudstones, siltstones and sandstones of the formation reflect deposition in a well-oxygenated, mainly quiet, shallow-marine environment. The change from poorly oxygenated conditions, which prevailed in the underlying Maentwrog Formation, occurred abruptly and, as there is no evidence of progressive shallowing, it is suggested that it reflects rapid eustatic, rather than local tectonic, changes. The dominance of structureless mudstone, probably homogenised by bioturbation, implies mostly quiet conditions, and many of the thinner siltstone beds are probably turbidites. Rare hummocky cross-stratification implies storm influence and water depths between fair-weather and storm wavebase, more than 15 m (Swift et al., 1983). Sharp bases and tool marks indicate that most siltstones and sandstones were emplaced initially by turbidity currents, but development of the full Bouma sequence was probably suppressed by storm waves (cf. Hamblin and Walker, 1979). The lenticular sandstone sequences near the top of the formation, which display proximal turbidite characteristics and probably fill channels, suggest deeper conditions, probably further offshore. Their derivation from the north contrasts with the bulk of the formation, in which palaeocurrents are mainly southerly derived (Crimes, 1970). Crimes described trace fossils of the *Cruziana* and *Skolithos* ichnofacies, respectively implying sublittoral and littoral conditions, from the *Lingulella*-rich beds near the top of the formation. The provenance of the quartz-dominated sediments was probably similar to that during Maentwrog Formation times.

Dolgellau Formation

The Dolgellau Formation comprises pyritous and carbonaceous black mudstone and crops out between Dolgellau and Tonfanau, except where it was removed by pre-Arenig erosion across the Dol-Ithel Anticline, north of Mynydd Moel (Figures 2 and 3). It commonly occupies a narrow depression and is poorly exposed. It is estimated to be about 100 m thick, but precise measurement is inhibited by numerous faults and intrusions at this horizon, particularly between the Gwynant valley and Dolgellau. Along much of the outcrop between Friog and the alluvium in the Mawddach estuary, the formation is juxtaposed by strike-parallel faults with the Ffestiniog Flags Formation. The base is rarely exposed, but near Garth Isaf [6601 1613; 6507 1546] it grades upwards from the Ffestiniog Flags Formation, and over about 10 m the colour changes from grey to black, with a gradual decrease in siltstone beds. Elsewhere, for example at Rhoslefain [5710 0536], the contact forms a distinctive dip slope, implying a sharper change in lithology.

The formation is well exposed beside the Mawddach estuary near Fegla Fach [6456 1562], but is intensely folded and faulted. A less-disrupted sequence is moderately well exposed in the Gwynant valley [6795 1705]. However, the best section is in the water-filled quarry [6210 1210], known locally as the 'Blue Lake', at Friog. The vertical north-east wall, and a higher bench to the east of the main excavation, expose about 80 m of monotonous black laminated mudstone with three or four beds of deeply weathered, calcareous cone-in-cone concretions up to 0.6 m long and 0.1 m thick. Bedding is defined by thin laminae, less than 1 mm thick, of disseminated pyrite and siltstone, and slight colour and grain-size variations, on weathered surfaces. Coatings of soft ochreous and iridescent material occur on cleavage, joint and bedding surfaces. A few phosphate nodules also occur. At another quarry [6247 1311], about 1 km north-east of Friog, similar mudstones are hornfelsed by a microdiorite intrusion.

BIOSTRATIGRAPHY

The lowest beds of the Dolgellau Formation yield the trilobite *Parabolinoides bucephalus* (Plate 2 g) by the Gwynant [6792 1705]; south of Gwern-y-barcud [6998 1754], slightly higher strata yield *Parabolina spinulosa* (Plate 2 f); both of these indicate the presence of the *P. spinulosa* Biozone. The same zone is indicated by the abundance of the brachiopod *Orusia lenticularis* with *Pseudagnostus* sp. and a fragment of *P. spinulosa*? in a quarry south-west of Rhoslefain [5700 0528]. At Bryn-rhug, Dolgellau, Lake and Reynolds (1896, p. 514) collected *O. lenticularis* and *P. spinulosa*, but the presence of *Leptoplastus* cannot be confirmed. Higher levels within the Dolgellau Formation have been determined only near Pont Abergwynant [6756 1707], where typical olenid faunas of the *Peltura scarabaeoides* Biozone are abundant but very poorly preserved. *Ctenopyge bisulcata, C. falcifera?, C. pecten, Lotagnostus sp.*, *Peltura* sp. and *Sphaerophthalmus humilis* suggest the presence of the *C. bisulcata* Subzone. The most southerly outcrop yielded *Sphaerophthalmus major* and this may represent the overlying *C. linnarssoni* Subzone.

CONDITIONS OF DEPOSITION

The formation reflects accumulation in poorly oxygenated bottom waters which probably resulted from the suppression of sediment input because of eustatic drowning of the source area. Apart from thin siltstone turbidites, sedimentation was by pelagic settling.

THREE

Ordovician

Within the district the Ordovician comprises a sequence of sedimentary and volcanic rocks about 5 km thick (Table 1; Figure 3). The basal Tremadoc Series incorporates the uppermost sedimentary rocks of the Mawddach Group and the Rhobell Volcanic Group. The overlying Aran Volcanic Group, of Arenig to Caradoc age, consists of basic and acid volcanic rocks with intercalated mudstones and is succeeded by a sequence of sedimentary rocks of Caradoc to Ashgill age. The volcanic rocks have been interpreted to reflect island arc (Rhobell Volcanic Group) and back-arc, extensional (Aran Volcanic Group) environments (Kokelaar et al., 1984). Sedimentation was dominantly of marine mudstones, with few turbiditic sandstones, and throughout the Tremadoc–Caradoc sequence there is much evidence of incorporation of volcaniclastic debris.

The Ordovician biostratigraphy is based mainly on graptolite biozones (Table 2). However, the sparse faunas and poor preservation inhibits precise subdivision (Rushton, 1990); the *murchisoni* and *teretiusculus* Biozones are particularly difficult to characterise and the problem is most pronounced in the middle of the Aran Volcanic Group, where ironstone development implies a probable non-sequence in the upper Llanvirn or Llandeilo (Table 1). Correlation between the graptolite biozones and the shelly stages is imprecise. New collections were made at several localities and, where available, certain older collections (e.g. by A H Cox) were re-examined. Microfloral studies have demonstrated that acritarchs are of particular value in the Tremadoc and Arenig series, but less so in post-Arenig strata (Molyneux, 1990).

Table 2 Ordovician biostratigraphy.

Series	Stages and zones (shelly fossils)	Graptolite zones
Ashgill	Hirnantian	*Normalograptus? persculptus*
	Rawtheyan	*Dicellograptus anceps*
	Cautleyan	*Dicellograptus complanatus*
	Pusgillian	
		Pleurograptus linearis
Caradoc	Onnian	
	Actonian	
	Marshbrookian	*Dicranograptus clingani*
	Woolstonian	
	Longvillian	
	Soudleyan	*Diplograptus multidens*
	Harnagian	
	Costonian	
		Nemagraptus gracilis
Llandeilo	—	*Glyptograptus teretiusculus*
Llanvirn	—	*Didymograptus murchisoni*
		Didymograptus artus
Arenig	Fennian	*Didymograptus hirundo*
		Isograptus gibberulus
	Whitlandian	*Didymograptus nitidus*
	Moridunian	*Didymograptus deflexus*
		?
Tremadoc	*Angelina sedgwickii*	
	Shumardia pusilla	
	?	?
		Adelograptus tenellus
		Rhabdinopora flabelliformis

THE MAWDDACH GROUP

The Dol-cyn-afon Formation is the uppermost formation of the Mawddach Group. The occurrence of the dendroid graptolite *Rhabdinopora [Dictyonema] flabelliformis* s.l. (Plate 2c) at its base distinguishes the base of the Ordovician System (Allen et al., 1981).

Dol-cyn-afon Formation

The Dol-cyn-afon Formation comprises mainly dark grey mudstone and was previously referred to as the Tremadoc Slates (Cox and Wells, 1921, 1927; Jones, 1933). Jones (1933) considered that on the west side of the Llanegryn Fault, the 'Tremadoc slates' were uplifted and removed by erosion. However, around Rhoslefain [5730 0520], up to 130 m of strata have been determined. The thickness increases eastwards to 350 m around Arthog and 475 m in the Gwynant valley, its maximum development. However, south of Dolgellau, around Mynydd-y-Gader, the formation appears to have been entirely removed by pre-Arenig erosion south of the Ceunant Fault (Figure 7). Cox and Wells (1921) subdivided the sequence into six units, based mainly on the contained faunas and very subtle changes in lithology, but they are difficult to sustain across the district.

At the 'Blue Lake' quarry [6210 1210], near Friog, the base of the formation is placed where black, pyritous mudstones of the Dolgellau Formation grade, over 2–3 m, into paler mudstones with abundant phosphate nodules. North-east of the Gwynant valley [6906 1719], the base lies at the contact between laminated dark grey mudstones and overlying paler mudstones. In a quarry at Arthog [6520 1521], the lowest 200 m of the formation comprises massive, well-cleaved, rusty weathered mudstones with traces of bioturbation. Bedding is poorly defined and can be distinguished only by widely spaced car-

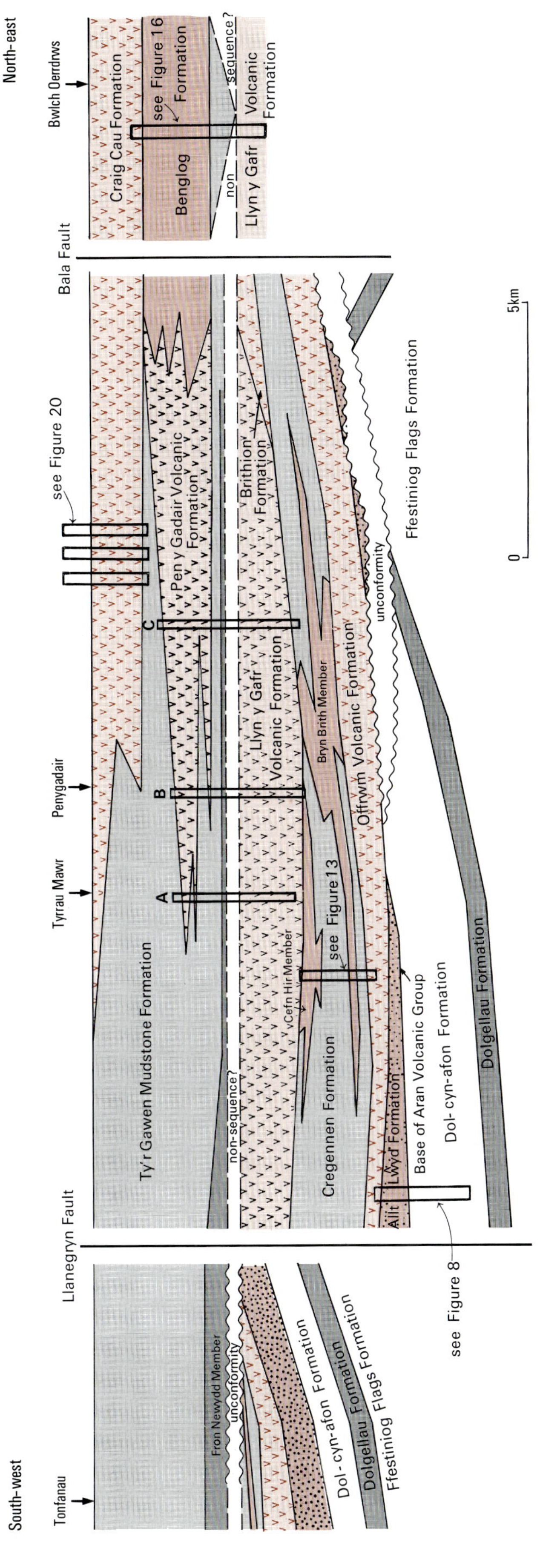

Figure 7 Simplified fence diagram of the Aran Volcanic Group. Columns A to C are depicted in more detail in Figure 15.

bonate concretions, pyrite laminae and layers of pyritised burrows and phosphate concretions.

At Garth Angharad [6684 1657], a massive sandstone, up to 10 m thick, in the lower part of the formation can be traced for about 250 m along strike. The fine-grained sandstone contains numerous, matrix-supported, angular mudstone clasts, up to 15 mm across.

Bedding is more clearly defined in the upper part of the formation, with a greater proportion of thin siltstones and sandstones. A sequence of massive, strongly bioturbated mudstone with streaks, clots and rare beds, less than 20 mm thick, of ripple cross-stratified, fine-grained sandstone and siltstone can be traced between Arthog and the Gwril valley. It is thickest, about 110 m, in the Gwril valley [6090 0931] where 5 m of coarse- to fine-grained, trough cross-bedded quartzose sandstone occurs close to its base (Figure 8). The sequence may represent the '*Asaphellus* Beds' (Cox and Wells, 1921) and it thins to about 50 m near Arthog [6498 1474], where it forms distinct dip-and-scarp topography.

Biostratigraphy

The early Tremadoc *Rhabdinopora flabelliformis* Biozone (Table 2) is recognised in the lower part of the Dol-cyn-afon Formation, especially around the Gwynant valley, but its subdivision into a *socialis* Subzone below, and a *flabelliformis* Subzone above, as in the Harlech district (Rushton, 1982), is not possible. Near Garth Angharad [6633 1631] and in the Gwynant valley [6805 1674], subspecies of the dendroid graptolite *Rhabdinopora flabelliformis*, typical of the 'Lower *Dictyonema* Band' (Cox and Wells, 1921), were determined (Plate 2c). Also, south-west of Craig y Castell [6881 1718], in the lower part of the formation (Cox and Wells, 1921, p.262), *Rhabdinopora flabelliformis flabelliformis* (sensu Bulman), *Eurytreta sabrinae*?, *Lingulella* and trilobite fragments, including *Niobella homfrayi*, are typical of the early Tremadoc.

North-east of the Gwynant Valley [6900 1703 to 6904 1719], lower Tremadoc acritarchs include *Acanthodiacrodium* cf. *ubui*, ?*Dasydiacrodium caudatum*, *Stelliferidium cortinulum*? and *S. simplex* (Plate 3). On the west side of the Gwynant valley [6794 1674], strata above the 'Lower *Dictyonema* Band' yielded ?*Cristallinium randomense* and ?*Veryhachium dumontii*, of possible early Tremadoc age. At Nant y Ceunant [7238 1574], in the faulted outcrop south of Dolgellau, *Acanthodiacrodium tumidum*, *A. angustum*, *Cymatiogalea cuvillieri* and *Vulcanisphaera africana* indicate the lower Tremadoc.

The *tenellus* Biozone has not been distinguished, but a diverse shelly fauna from slate trials at Llyn Wylfa [6699 1630], south-east of Garth Angharad, which includes '*Conularia*' *homfrayi*, *Eurytreta sabrinae*?, *Peelerophon*? *arfonensis*, *Lingulocaris lingulaecomes*?, hyolithids and the trilobites *Asaphellus homfrayi* (Plate 2b), *Leptoplastides* sp., *Micragnostus calvus*, *Orometopus praenuntius*, *Pseudokainella*

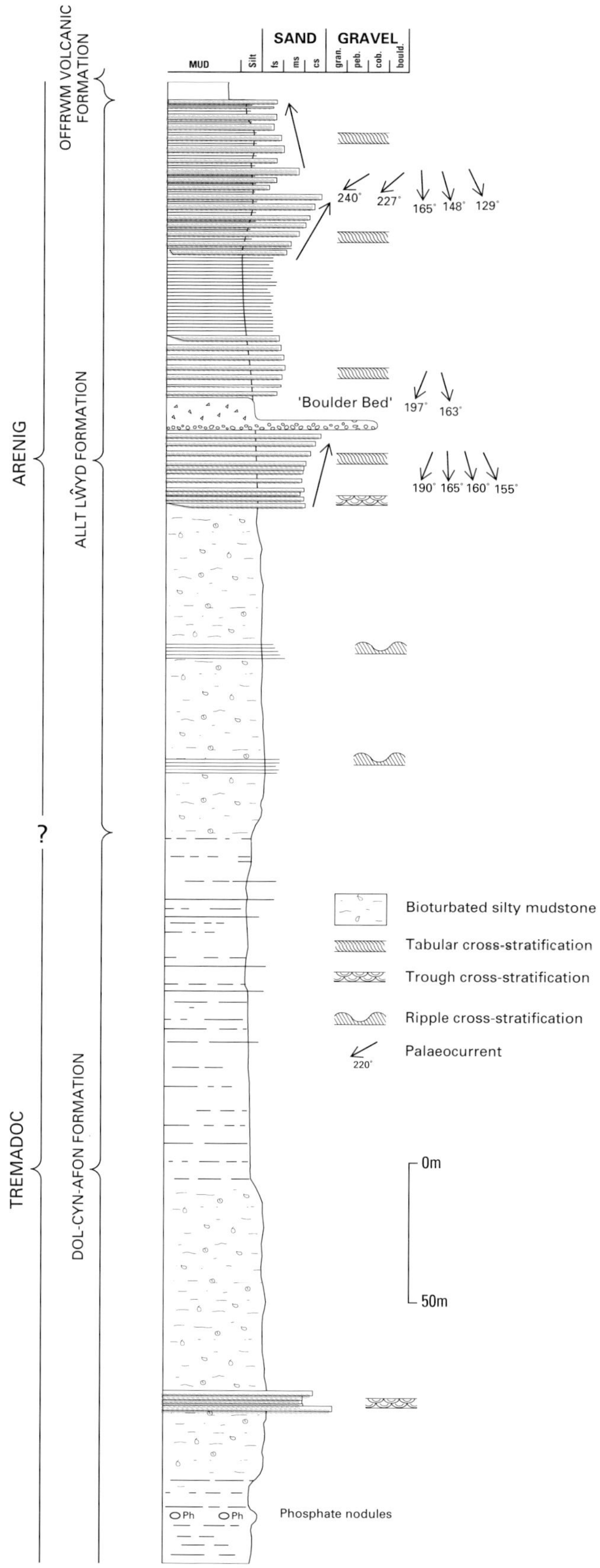

Figure 8 Sedimentary log of the Allt Lŵyd Formation and upper part of the Dol-cyn-afon Formation in the Gwril valley [6080 0900] (see Figure 9). Upward-fining and upward-coarsening sequences are indicated by large arrows.

impar (Plate 2a), *Shumardia* (*Conophrys*) sp., clearly indicates the upper Tremadoc *pusilla* Biozone. Above this horizon, Cox and Wells (1921) described the 'Upper *Dictyonema* Band' [6818 1652] which has yielded an abundance of close-meshed *Rhabdinopora* resembling *R. flabelliformis socialis*, and one example of *R. f.* cf. *anglica* (Plate 2d). Acritarchs include *Acanthodiacrodium angustum, A. ovatum, A. tuberatum, Cymatiogalea cristata, C. multarea,* ?*C. velifera,* ?*Stelliferidium fimbrium, S. stelligerum, Vulcanisphaera africana, V. britannica* and *V. frequens*, and indicate a late Tremadoc age, probably *pusilla* Biozone (Plate 3). Although the specimens of *Rhabdinopora* are more characteristic of the earlier Tremadoc, a similar anomalous association with *pusilla* Biozone trilobites and acritarchs was reported from the Deanshanger Borehole of Buckinghamshire (Bulman and Rushton, 1973).

Elsewhere within the district, middle and upper Tremadoc strata have been determined on acritarch evidence. In the Gwril Valley [6089 0938 and 6088 0935] *Acanthodiacrodium ovatum*? and *Stelliferidium distinctum*? suggest a stratigraphical level equivalent to the *pusilla* Biozone in the Shineton Shales. East of the Bala Fault, near Cae'r-tyddyn [7840 1778], *A. angustum, C. cristata, C. cuvillieri, S. fimbrium*? and *V. frequens* suggest a possible late Tremadoc age.

Conditions of deposition

The formation was deposited under relatively quiet conditions, which encouraged bioturbation and facilitated pyrite and phosphate crystallisation. Bioturbation is clearly distinguished only in the possible '*Asaphellus* Beds', but the ill-defined bedding in much of the mudstone sequence was also probably caused by bioturbation. West of the Llanegryn Fault the lower part of the Allt Lŵyd Formation, which is of Tremadoc age, may be equivalent to the '*Asaphellus* Beds'. It is possible that these facies changes reflect syndepositional activity on the fault (cf. Jones, 1993; Kokelaar, 1988), but it is difficult to be certain because of the considerable displacement during Mesozoic/Tertiary times.

THE RHOBELL VOLCANIC GROUP

The Rhobell Volcanic Group overlies the Mawddach Group around the south-eastern side of the Harlech Dome and wedges out within the district (Allen and Jackson, 1985; Figure 3; Table 1). It crops out only in a small wedge west of Dolgellau and comprises grey, porphyritic, locally autobrecciated basaltic lavas. Near Maes Angharad [7077 1776], two feldsparphyric, basaltic lavas, 3 and 9 m thick, are intercalated with basaltic crystal tuffs.

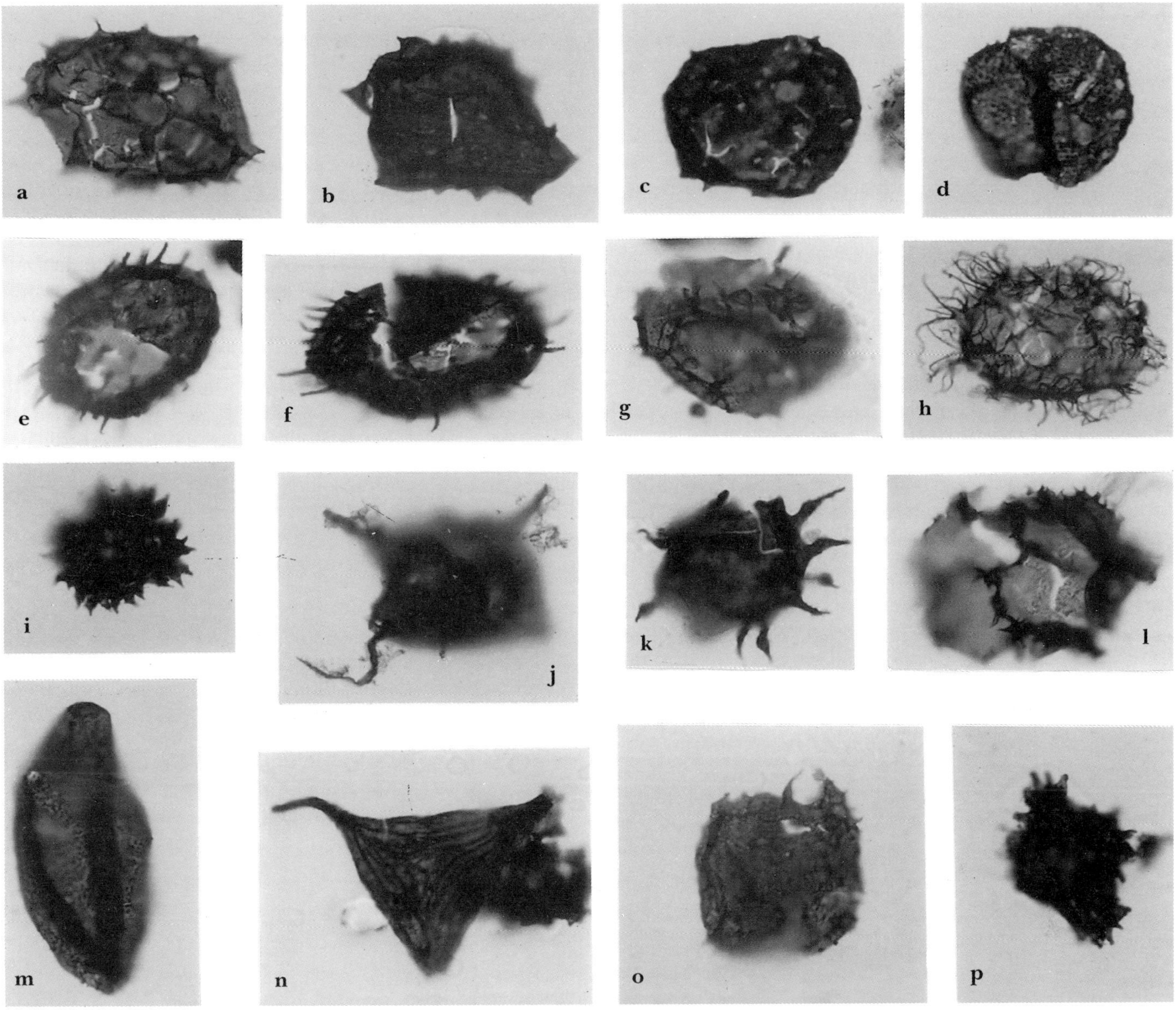

Plate 3 Acritarchs from the Dol-cyn-afon and Allt Lŵyd formations. Specimens are housed in the MPK collection of the Biostratigraphy and Sedimentology Group, BGS, Keyworth. All specimens ×1000.

a) *Acanthodiacrodium ovatum* Rasul. MPK 9725. Sample number MPA 33564, 'Upper *Dictyonema* Band' of Cox & Wells (1921), Dol-cyn-afon Formation, Gwynant valley [6818 1652].
b) *Acanthodiacrodium tuberatum* (Downie) Martin. MPK 9726. Sample number MPA 33564.
c) *Acanthodiacrodium tumidum* (Deunff) Martin. 1977. MPK 9727. Sample number MPA 37965, Dol-cyn-afon Formation, Nant y Ceunant [7238 1574].
d) *Acanthodiacrodium angustum* (Downie) Combaz. MPK 9728. Sample number MPA 37965.
e) *Stelliferidium simplex* (Deunff) Deunff, Górka & Rauscher. MPK 9729. Sample number MPA 37984, Dol-cyn-afon Formation, Dolgedr [6899 1719].
f) *Stelliferidium simplex*. MPK 9730. Sample number MPA 37984.
g) *Cymatiogalea cristata* (Downie) Rauscher. MPK 9731. Sample number MPA 33564.
h) *Vulcanisphaera cirrita* Rasul. MPK 9732. Sample number MPA 33565, 'Upper *Dictyonema* Band', Dol-cyn-afon Formation, Gwynant valley [6818 1652].
i) *Micrhystridium* aff. *acuminosum* Cramer & Diéz. MPK 9734. Sample number MPA 33594, Allt Lŵyd Formation, near Tydyyn-Meurig [5745 0462].
j) ?*Marrocanium simplex* Cramer et al. MPK 9734. Sample number MPA 33570, Allt Lŵyd Formation, south of Llyn Wylfa [6749 1624].
k) *Stellechinatum sicaforme* Molyneux *in* Molyneux & Rushton (1988). MPK 9735. Sample number MPA 33594.
l) *Cristallinium randomense* Martin. MPK 9736. Sample number MPA 33593, Allt Lŵyd Formation, near Tyddyn-Meurig [5742 0468].
m) *Dasydorus cirritus*? Playford & Martin. MPK 9737. Sample number MPA 33571, Allt Lŵyd Formation, south of Llyn Wylfa [6749 1623].
n) *Arkonia virgata* Burmann. MPK 9738. Sample number MPA 34997, Allt Lŵyd Formation, east of Cross Foxes [7725 1671].)
o) *Coryphidium* sp. MPK 9739. Sample number MPA 33571 (as fig. m)
p) *Coryphidium* aff. *bohemicum* Vavrdová. MPK 9740. Sample number MPA 34997 (as fig. n).

THE ARAN VOLCANIC GROUP

The Aran Volcanic Group (Arenig–Caradoc) comprises 2–2.5 km of volcanic and sedimentary rocks, which overlie, in places unconformably, the Mawddach and Rhobell Volcanic groups. It crops out between Tonfanau, in the west, and Bwlch Oerddrws, in the east, and over much of this outcrop two acid tuff formations and two basalt formations, with intercalated mudstones, can be distinguished (Figure 7). Most of the volcanic rocks wedge out westwards and to the east the basalts pass into a thicker, mainly volcaniclastic sequence. The associated mudstones are commonly contaminated with volcanic detritus.

Allt Lŵyd Formation

The Allt Lŵyd Formation (Allen and Jackson, 1985), previously the Basement Series/Beds/Group (Cox, 1925; Cox and Wells, 1921, 1927; Jones, 1933), comprises shallow-marine sandstones, conglomerates and silty mudstones. It is thickest, up to 290 m, between Tonfanau and the Gwril valley, but it thins eastwards to 180 m at Arthog, 60 m near Pant Phylip [6476 1403], 10 m south of Garth Angharad [6734 1635], and wedges out near the Gwynant valley (Figure 9). The formation is absent across much of Mynydd-y-gader, south-west of Dolgellau, but is thinly developed south of Dolgellau and in the north-east corner of the district, about the Bala Fault. South-west of Dolgellau, an estimated 200 m of strata occurs in a fault-bound area north of the Ceunant Fault.

Over much of the outcrop, the contact with the underlying Dol-cyn-afon Formation is transitional and conformable, but in places, south-west of Dolgellau, the formation lies unconformably on the Ffestiniog Flags Formation (Figure 7). In the Gwril valley [6110 0880] and near Arthog [6308 1319], the base is placed where thin siltstones and sandstones become dominant over silty mudstones (Figure 8). Elsewhere, as near Fairbourne [6301 1304] and Rhoslefain [5736 0513], the base is sharp and marked by 8–15 m of massive, coarse-grained, quartzose sandstones which closely resemble the Garth Grit of North Wales (Jennings and Williams, 1891; Nicholas, 1915; Allen and Jackson, 1985; Figure 9). The sandstones are ill sorted with scattered quartz pebbles in structureless beds up to 2.5 m thick.

Above the basal sandstones, the formation comprises either thinly bedded, flaggy, fine-grained sandstone and silty mudstone, probably equivalent to the 'Llyfnant Flags' of Fearnsides (1905), or medium- to massive-bedded sandstones. The flaggy strata dominate the western outcrops but are thin and locally absent in the east of the district (Figure 9).

Between Tonfanau and Rhoslefain [5813 0619], 5–15 m-thick sequences of medium- to thick-bedded sandstone, which form ridges, are separated by 30–60 m-thick sequences of thinly bedded, flaggy sandstones and

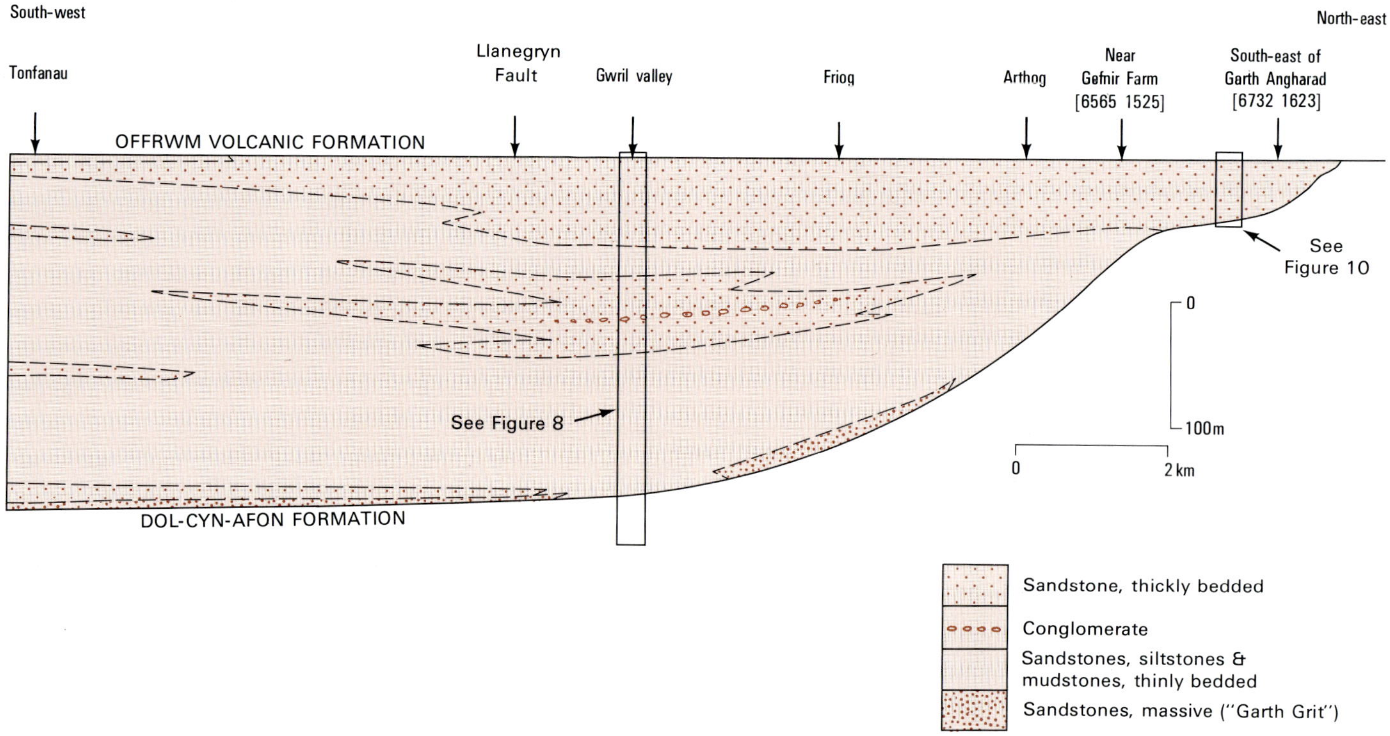

Figure 9 Generalised lithological variations within the Allt Lŵyd Formation between the coast and Garth Angharad. Columns indicate the position of the detailed sedimentary logs of Figures 8 and 10.

silty mudstones. The thick sandstone beds are normally graded, massive, planar laminated or display tabular cross-stratification. At the south end of the ridge [5801 0582], a lenticular bed, more than 3 m thick, of medium-grained, structureless sandstone with scattered pebbles of subrounded quartz up to 5 mm in diameter, possibly represents a channel fill. The flaggy sandstones, 1–10 mm thick, are laterally discontinuous with parallel-lamination and starved, cross-stratified ripples. Loaded bases and *Chondrites*-type bioturbation are common.

In the Gwril valley, the formation comprises 120 m of flaggy, disrupted sandstones and mudstones overlain by 150 m of dominantly thickly bedded, quartzose and feldspathic sandstone (Figure 8). Despite widespread vertical, *Skolithos*-type bioturbation, planar lamination and tabular and trough cross-stratification are locally preserved. The coarsest sandstones contain rip-up clasts of mudstone and subangular to subrounded pebbles of quartz and acid volcanic rock, up to 10 mm in diameter, as lags along foresets or the bases of beds. Foreset attitudes and localised current drag folding of foresets indicate mainly southerly directed palaeocurrents (Figure 8).

At Llyn Wylfa [6730 1625], south-east of Garth Angharad, the formation, 60 m thick, is entirely of sandstone beds, 0.1–2.5 m thick, with thin interbeds of siltstone and mudstone. Four upward-thickening and upward-coarsening cycles, 2–15 m thick, can be distinguished (Figure 10). The sandstones are generally moderately to well sorted, fine to coarse grained. Similar sandstones are exposed near Dolgellau [7277 1729 and 7260 1720], and along Afon Arran [7299 1725 and 7295 1735].

The Aran Boulder Bed, which is up to 300 m thick in the Bala and Harlech districts (Dunkley, 1978, 1979; Allen and Jackson, 1985), wedges out near Cae'r-tyddyn [7894 1776], in the north-east of the district. It is represented by 4 m of poorly sorted conglomerate, with subrounded pebbles of volcanic rocks up to 25 mm in diameter, at, or close to, the top of the formation. No conglomerates are distinguished between Cae'r-tyddyn and Arthog, but between the Llanegryn Fault and Arthog, a thin conglomerate is possibly an equivalent bed. It is 11 m thick south-east of Friog [6222 1162 and 6215 1141] and overlies 6 m of sandstones which coarsen and thicken upwards. The well-rounded, clast-supported pebbles, which are mainly 50–100 mm in diameter, but locally up to 0.4 m, are normally graded in the upper few metres of the bed. They are mainly of andesite or basaltic andesite composition, with a few of laminated siltstone. The sparse matrix is of cleaved tuffaceous mudstone and layers, up to 50 mm thick, with fewer clasts distinguish poorly defined bedding. In the Gwril valley, the bed is 13 m thick, but the andesite pebbles are smaller, 20–60 mm across, matrix supported and confined to the basal 2 m. Above, the bed comprises cleaved tuffaceous mudstone with thin layers of angular, acid volcanic rock fragments (Figure 8).

Petrography

The massive basal sandstones are generally quartz rich. Near Fairbourne [6240 1219], the sandstones consist of moderately sorted, subangular to subrounded quartz grains, up to 6 mm diameter, with minor amounts of

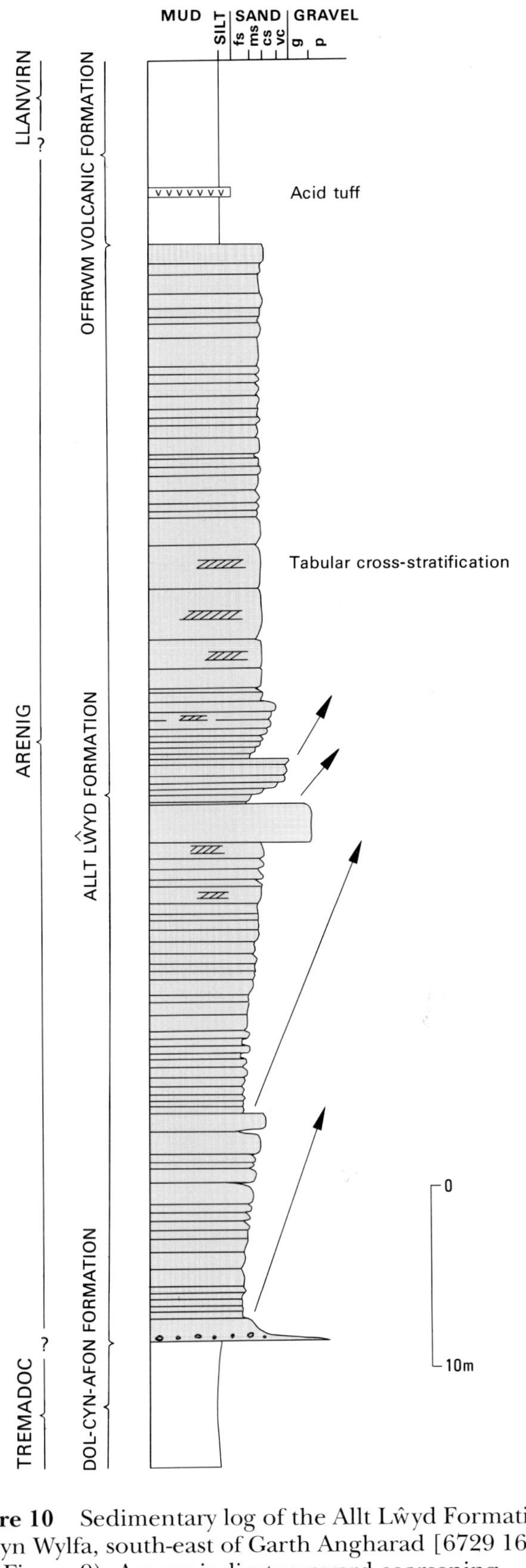

Figure 10 Sedimentary log of the Allt Lŵyd Formation at Llyn Wylfa, south-east of Garth Angharad [6729 1626] (see Figure 9). Arrows indicate upward-coarsening sequences.

feldspar, mica and opaque/heavy minerals (Plate 5a). Quartz grains and quartzose rock fragments make up about 80 per cent of the rock.

The thin flaggy sandstones, such as those near Gefnir Farm, are generally dominated by angular grains of sericitised plagioclase, quartz, some strained, and minor detrital white mica (E62029). Chlorite-mica stacks, commonly concentrated at the tops of individual sandstone beds, are also locally abundant. The thickly bedded sandstones are variably quartz-, feldspar- and lithic-rich, moderately to well sorted, and, with the exception of scattered, rounded quartz pebbles, contain mainly angular to subangular grains. In the Gwril valley [6117 0883], sandstones in the upper part of the formation contain up to 40 per cent feldspar and some basalt fragments (E65805). Lithic clasts and feldspars are more common in sandstones in the east part of the outcrop (e.g. E62037, E65934) (Plate 5b). The feldspars are commonly euhedral, up to 2 mm in length, and are locally altered to calcite and white mica (E62029–31). Lithic clasts are mostly of feldsparphyric, flow-foliated basalt (E65584), highly vesicular basalt (E651230) and fine-grained acid volcanic rocks (E62030).

The andesite, or basaltic andesite, pebbles of the conglomerate contain euhedral, strongly zoned feldspar phenocrysts in a devitrified glassy or microcrystalline groundmass (Plate 5c). The matrix comprises weakly cleaved, chloritic mudstone with fragments of chloritised glass, andesite and highly altered, clinopyroxene-bearing basalt (E65583–4).

Biostratigraphy

West of the Llanegryn Fault, between Tonfanau and Rhoslefain [5736 0479, 5742 0468], the lower part of the formation has yielded early Tremadoc acritarchs including *Cymatiogalea cuvillieri, Cristallinium randomense, Stelliferidium* cf. *simplex, Vulcanisphaera africana* and *V. cirrita* (Plate 3). Nearby [5745 0462, 5700 0393], the upper part of the formation has yielded early Arenig acritarchs, including *Micrhystridium* aff. *acuminosum*, abundant and varied *Polygonium* spp., *Stellechinatum sicaforme*, and *Striatotheca prolixa*, suggesting a pre-*deflexus* Biozone age and a possible depositional break (Table 2; Molyneux and Rushton, 1988; Molyneux and Dorning, 1989; Cooper and Molyneux, 1990). Near the top of the formation at Tonfanau [5744 0451], a flora with abundant and varied *Polygonium* spp., *Veryhachium lairdii* and *V. trispinosum* also suggests a possible early Arenig age.

South-east of Garth Angharad, floras from the lower part of the formation [6749 1624, 6749 1623, 6716 1621] comprise *Coryphidium* cf. *bohemicum, C. minutum*?, *Micrhystridium* aff. *acuminosum*, ?*Marrocanium simplex, Polygonium* spp., *Stellechinatum* cf. *uncinatum, Stelliferidium* spp. and *Veryhachium trispinosum*, and suggest an early Arenig (Moridunian) age (Table 2; Plate 3). In the upper part [6749 1622], the association of *Dasydorus cirritus*?, ?*Striatotheca rarirrugulata, Veryhachium lairdii* and abundant *V. trispinosum* suggests a late Arenig (Fennian) age, equivalent to the *gibberulus* or *hirundo* Biozone (Table 2; Molyneux, 1987, 1990).

East of Cross Foxes [7879 1759, 7892 1785, 7900 1782], the recognition of *Coryphidium*? and *Veryhachium trispinosum*? suggests a possible Arenig age. Diverse floras from a stream section east of Gwanas Fawr [7720 1670] include *Coryphidium*? sp., *Polygonium* sp., *Stelliferidium* sp., *Striatotheca principalis parva, Veryhachium* aff. *lairdii* and *Veryhachium* aff. *trispinosum*; the *V.* aff. *lairdii* suggests a possible late Arenig age (Rushton and Molyneux, 1989). In the vicinity [7725 1671], a diverse flora includes *Arbusculidium filamentosum, Arkonia virgata, Coryphidium* aff. *bohemicum, Frankea hamata, F. sartbernardensis, Marrocanium simplex, Stellechinatum celestum, Striatotheca quieta*?, *S. rarirrugulata*?, *Veryhachium lairdii* and *V. trispinosum* (Plate 3). The presence of *A. virgata, S. celestum* and *S. quieta*? indicates a Llanvirn or younger age, but *C.* aff. *bohemicum* and *S. rarirrugulata*? recall late Arenig assemblages, suggesting that a Llanvirn age is more probable. In the eastern part of the stream section [7732 1666], the flora includes *Striatotheca* sp. cf. *S. frequens, S. principalis, S. rarirrugulata, Veryhachium lairdii* and *V. trispinosum*, suggesting a late Arenig or younger age.

Conditions of deposition

The shallow-water bedforms, *Skolithos*-type burrows, rounded pebbles and sorting of the thickly bedded sandstones suggests that they were deposited in shallow marine conditions. The palaeocurrent data from the Gwril valley (Figure 8), which indicate transport from a single direction, implies a storm, rather than a tidally influenced, regime. The tabular cross-stratified beds probably represent straight-crested bars or sandwaves and the upward-coarsening and upward-thickening sequences indicate nearshore storm deposits. The strongly bioturbated, flaggy sandstones and mudstones are interpreted to reflect slightly deeper, quieter conditions, although within wave influence, but it is possible that they may represent intertidal mud-flat deposition. The conglomerates represent debris-flow deposits.

Within the formation, the diachronism, thickness and broad facies changes are intepreted to represent the progressive eastward migration of shallow-water sandstones on to an uplifted area in the north-east of the district. In the west, flaggy sandstone deposition began in Tremadoc times and a thick sequence accumulated. In the north-east, uplift and erosion, possibly contemporaneous with Rhobell Volcanic Group activity, exposed the Ffestiniog Flags Formation and the area was encroached with sandstone deposition during early Arenig times. The form of the uplifted block cannot be distinguished because of the restricted outcrop.

The volcanic component, basaltic andesite pebbles and euhedral feldspars, of the sandstones and conglomerates, and the increasing volcanic contamination towards the north-east, suggests contemporaneous erosion of the Rhobell Volcanic Group (Kokelaar, 1979). However, the dominant quartz component, in grains and pebbles, suggests derivation from an igneous/metamorphic landmass or, possibly, recycling of coarse sandstone from an uplifted area of the Harlech Grits Group to the north (Crimes, 1970; Bates, 1972; Beckly, 1987; Traynor, 1990).

Offrwm Volcanic Formation

The Offrwm Volcanic Formation (Allen and Jackson, 1985), up to 290 m thick, comprises acid tuffs, thin tuffites and mudstones. The strata were previously referred to as the Lower Acid, Mynydd-y-gader or Beacon Hill Acid groups (Cox, 1925; Cox and Wells, 1921, 1927; Jones 1933). In the west of the district the formation conformably overlies the Allt Lŵyd Formation, but eastwards it oversteps on to the Dol-cyn-afon Formation, near Garth Angharad, and the Dolgellau and Ffestiniog Flags formations, south of Dolgellau (Figure 7).

Acid tuffs are the main element of the sequence around Tonfanau, in the west, and Mynydd-y-gader, in the east, but in much of the intervening area the sequence is dominantly of mudstone (Figure 11). The formation thins eastwards from about 290 m at Tonfanau, to less than 50 m near Bodowen [5857 0550]. At Llwyngwril, west of the Llanegryn Fault [5948 0878], the formation is overstepped by the Fron Newydd Member (Figure 7). Correlation of individual tuffs is difficult, particularly in the poorly exposed ground around the Llanegryn Fault, but broadly they can be subdivided into crystal-rich and crystal-poor varieties.

At Tonfanau [5745 0345], the formation comprises sequences of tuffs, up to 80 m thick, separated by dark grey mudstones (Figure 11). The pale weathered tuffs are massive and contain numerous quartz, and fewer albite, crystals (Plate 5d). Generally, a weak bedding-parallel foliation is defined by undulating, iron-stained surfaces,

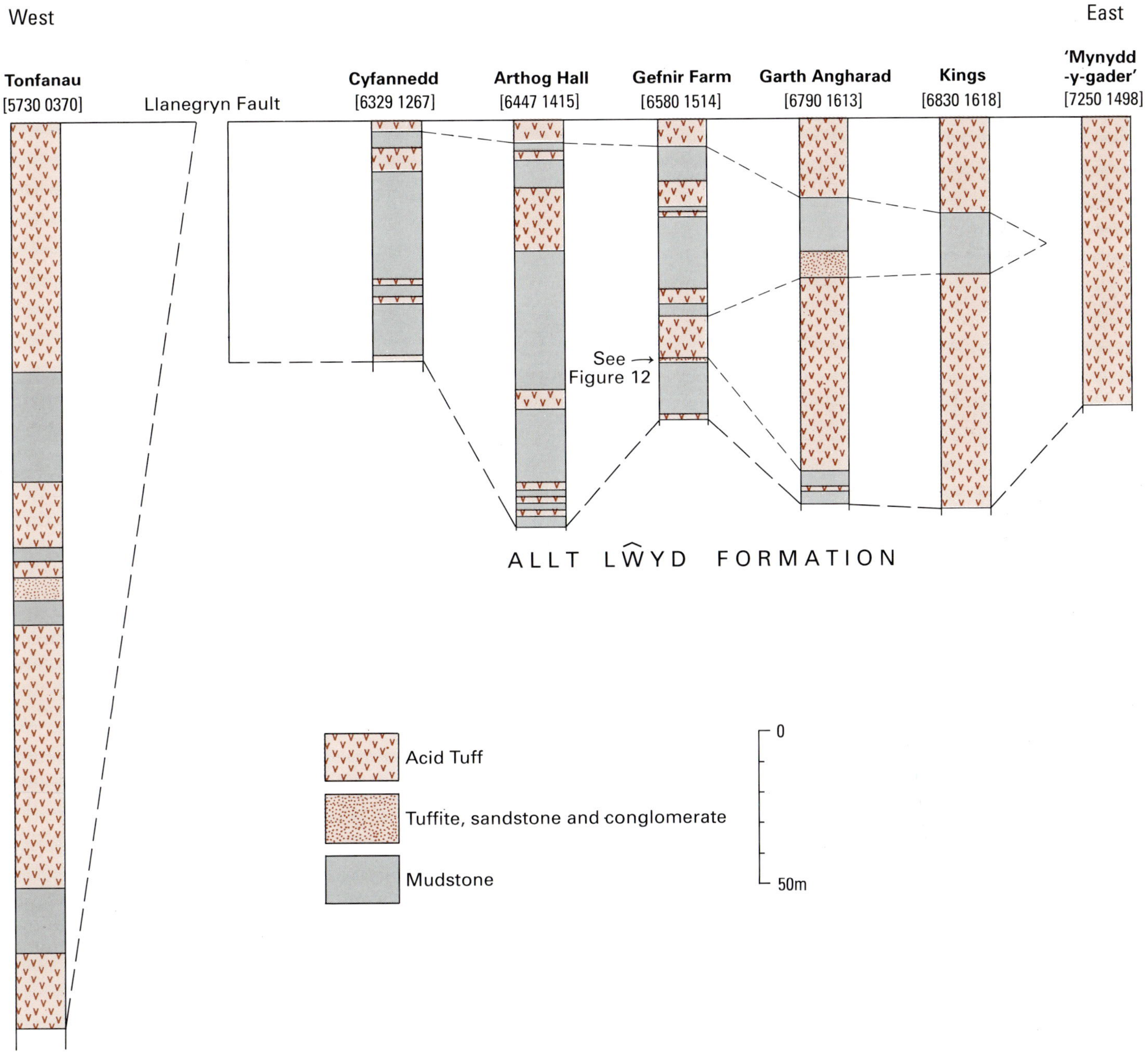

Figure 11 Lithological variations within the Offrwm Volcanic Formation.

spaced 0.5–1 m apart, but locally, as in a small quarry [5720 0307], the tuffs are strongly foliated. The groundmass of the tuffs (E67055-7) comprises a devitrified, recrystallised quartzofeldspathic aggregate but, in places, nonwelded and welded vitroclastic fabrics and pumice fragments can be distinguished. The euhedral albite crystals, up to 3 mm long, are variably altered with inclusions of sericite, calcite and epidote (E67087). Both feldspar and quartz crystals are commonly resorbed and embayed.

At Tonfanau, the intercalated strata are generally poorly exposed, but in a quarry [5727 0357] mudstones with 2 m of thickly bedded tuffite, comprising angular fragments of acid tuff and euhedral quartz crystals within a mudstone matrix (E67089) are well exposed. Nearby [5759 0402], a similar tuffite includes fragments of vesiculated basalt and flow-foliated microcrystalline basalt (E67078) and associated thin sandstones, up to 70 mm thick, are rich in quartz, detrital muscovite and fragments of acid tuff (E67071). Muddy tuffite beds are also exposed near Bodowen [5857 0550, 5858 0556], approximately 2 km to the north-east of Tonfanau.

In the vicinity of Arthog, and particularly near Gefnir Farm [6588 1534], the formation, 70–130 m thick, is well exposed and comprises mudstones with subordinate acid tuffs up to 20 m thick (Figures 11, 12). The basal beds are mainly mudstones with acid tuff beds, up to 2 m thick, but, south of Garth Angharad [6644 1575], they comprise tuffaceous mudstones, 2 m thick, with angular clasts, up to 60 mm across, of acid tuff and silty mudstone, feldspar crystals and sponge spicules (E62040). West of Gefnir Farm [6539 1484] the Allt Lŵyd Formation is conformably overlain by 2 m of crystal-rich acid tuff, with 1.5 m of thinly bedded, fine-grained acid tuff above.

At Arthog, the overlying tuffs are mainly massive and pale weathered, with a weak, bedding-parallel foliation accentuated by recrystallisation. Most tuffs contain fragments of acid tuff and, more rarely, basalt. For example, at Cyfannedd [6329 1267], angular fragments up to 0.15 m across are concentrated low in an 8 m-thick, massive tuff. Also, at Gefnir, subrounded, acid tuff fragments, up to 50 mm long, and mudstone fragments are common in a matrix of devitrified and recrystallised shards and tubular pumice (E66783, E66785). Fragments of granophyric intergrowths of quartz and feldspar have also been determined in some tuffs (E62032).

Most massive tuffs are overlain, and less commonly underlain, by bedded tuffs, in sequences, 0.15–1 m thick. These are either thick, normally graded and trough cross-stratified beds of tuff or thinly bedded, very fine grained and flinty, with common ripple cross-stratification. The thinly bedded tuffs are devitrified dust tuffs, the 'Chinastones' of Cox and Wells (1921), and locally, as east of Arthog [6444 1405], they form discrete sequences, up to 6 m thick, within mudstone.

Some tuffs contain a strong, bedding-parallel welding foliation which, in places, as at Gefnir [6581 1512], is deflected around large siliceous nodules. At both Gefnir and Arthog, the uppermost acid tuff, 5–6 m thick, is welded with a basal layer, 0.5 m thick, crowded with nonwelded pumice lapilli, 5–10 mm long (E66779, E66782, E66785).

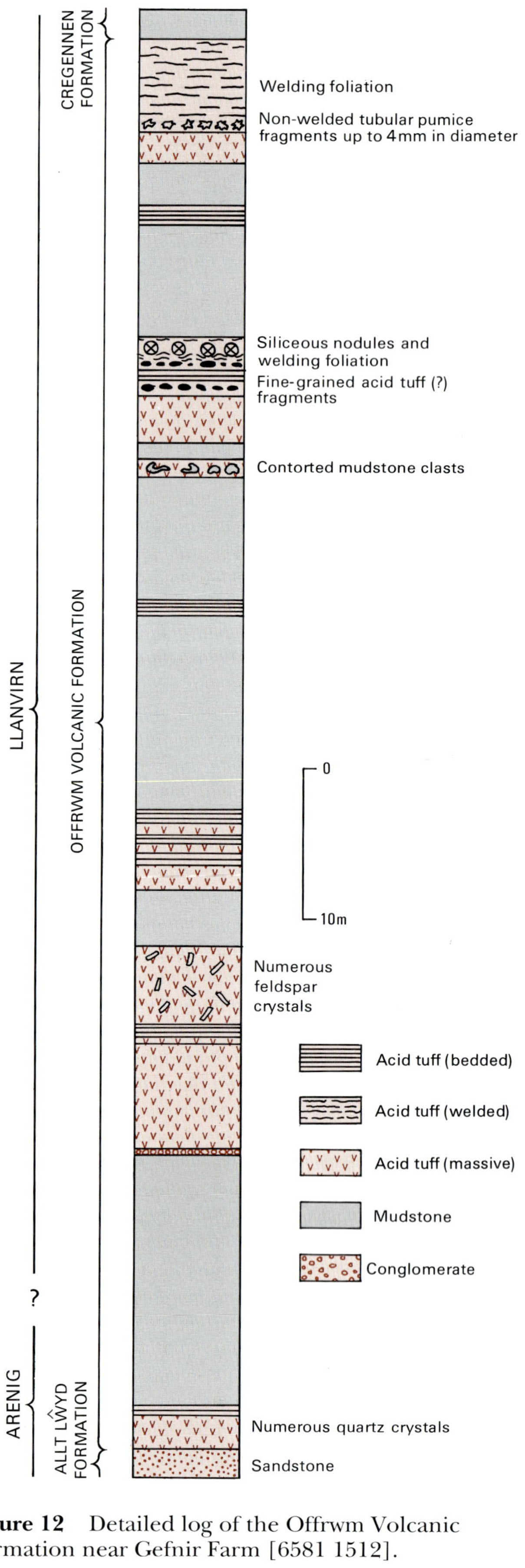

Figure 12 Detailed log of the Offrwm Volcanic Formation near Gefnir Farm [6581 1512].

East of Gefnir the massive tuffs thicken and around Kings [6834 1622] the formation comprises lower and upper tuff sequences, 75 m and 30 m thick respectively, separated by 20 m of mudstone (Figure 11). Farther east, on Mynydd-y-gader, the formation, up to 150 m thick, is mainly of acid, ash-flow tuff. The base is exposed in places on the north side of the dolerite intrusion [7245 1570] and comprises massive, lapilli-rich acid tuff, about 4.5 m thick, with bedded tuff, 1.5 m thick, above. The clasts, of pale grey tuffaceous mudstone (E65937), are subrounded, up to 0.3 m in diameter, and reverse graded in the basal 1 m, but normally graded above. The matrix is vitroclastic and non welded, with well-defined shards and pumice fragments (E65936). Above, the tuffs contain numerous quartz crystals and few intercalated bedded tuffs have been distinguished. Mudstones have been distinguished only between Kings and Gwernan Lake [e.g. 6983 1600].

South of the Mynydd-y-gader dolerite [7364 1520 to 7310 1493], the formation, 80 m thick, consists of pale grey weathered, massive, acid, ash-flow tuff. In places [7308 1501], an apparent well developed foliation is most probably due to siliceous recrystallisation (Plate 4a) although highly recrystallised, welded vitroclastic

Plate 4a Acid tuff with welding enhanced by silicification and recrystallisation. Offrwm Volcanic Formation. Mynydd-y-gader [7308 1501].

Plate 4b Basaltic debris-flow deposit. Cregennen Formation. Mynydd-y-gader [7233 1480].

Plate 4c Llynnau Cregennen and Pared y Cefn-hir viewed from the south-west. The prominent ridge is formed by the Cefn Hir Member and the rocky slopes to the right are in microgranite.

textures have been determined (E65943, E65949). Embayed subhedral quartz crystals are also common (E65950). Thinly bedded, fine-grained tuffs, up to about 3 m thick, occur locally [7257 1490] at the top of the formation.

BIOSTRATIGRAPHY

South-west of Beacon Hill, Tonfanau [5735 0350], mudstones in the Offrwm Volcanic Formation yielded *Didymograptus* (*D.*) *artus*? and *Barrandia* sp. juv., suggestive of the *artus* Biozone (Table 2). Localities near Bryn Brith [6664 1581 to 6642 1563] yielded poorly preserved graptolites, with some from the collections of A J Beckly and A H Cox, including '*Amplexograptus*' *confertus*, *Didymograptus* (*D.*) cf. *miserabilis*, *Eoglyptograptus* cf. *dentatus* and *Pseudoclimacograptus* sp. These are probably of lower Llanvirn, *D. artus* Biozone age (Fortey et al., 1990). Among similar faunas collected from north of Kings [6834 1622], Cox and Wells (1921, p. 272) identified "*Didymograptus nitidus*" and on that basis considered that the strata lie near the Arenig–Llanvirn boundary. However, re-examination of the Cox collection (in National Musuem of Wales) has not supported the determination of *D. nitidus*. Furthermore, specimens identified as *D. nitidus* do not range as high as the top of the *hirundo* Biozone in the Skiddaw Group succession (Fortey et al., 1990), so the identification is considered suspect.

CONDITIONS OF DEPOSITION

The acid tuffs were erupted following uplift and possible emergence in the north-east of the district. The absence of coarse terrigenous sedimentary rocks indicates marked subsidence. However, around Mynydd-y-gader, the thick accumulation of acid tuffs and unconformable base may reflect their subaerial eruption. On the basis of the thick accumulation of tuffs at Tonfanau, Jones (1933) and Kokelaar (1988) suggested the volcanic centre lay to the west of the district. At Mynydd-y-gader, the thick sequence of petrographically similar tuffs suggests the same source. However, the absence of thick tuffs between Arthog and Llwyngwril is difficult to explain. The lack of shallow water sedimentary structures and the thick intercalated marine mudstones indicate that the ash-flow tuffs were subaqueously emplaced, and some retained sufficient heat to weld. The thinly bedded tuffs, tuffites and tuffaceous sandstones indicate reworking of pyroclastic debris into debris flows and turbidity currents.

Cregennen Formation

The Cregennen Formation, up to 350 m thick, comprises dark grey mudstones, which are generally poorly exposed, with impersistent basic tuffs and tuffites and acid tuffs. The mudstones were previously referred to as the Moelyn, Crogenen and Bifidus slates (Cox and Wells, 1921, 1927; Jones, 1933). The formation conformably overlies the Offrwm Volcanic Formation in much of the outcrop, but south of Llwyngwril it is transgressed by the eastward overstep of the Fron Newydd Member (Figure 7). The volcaniclastic rocks, formerly the Bryn Brith Beds and Cefn Hir Ashes (Cox and Wells, 1921, 1927), now members, are clearly defined between Arthog and Llynnau Cregennen, where they form the ridges of Bryn Brith and Pared y Cefn-hir, but elsewhere their correlation is uncertain. Eastwards, the members merge and cannot be distinguished in the thick volcaniclastic sequence east of the Ceunant Fault, around Mynydd-y-gader. To the west, the members thin and wedge out. In the west of the district, near Tonfanau [5773 0371], the formation includes a single bed of massive, poorly sorted, basic tuffite, 8–10 m thick, which comprises vesicular and feldspar-phyric basalt, siltstone, mudstone and porphyritic rhyolite fragments in a mudstone matrix, rich in plagioclase and embayed, euhedral quartz crystals.

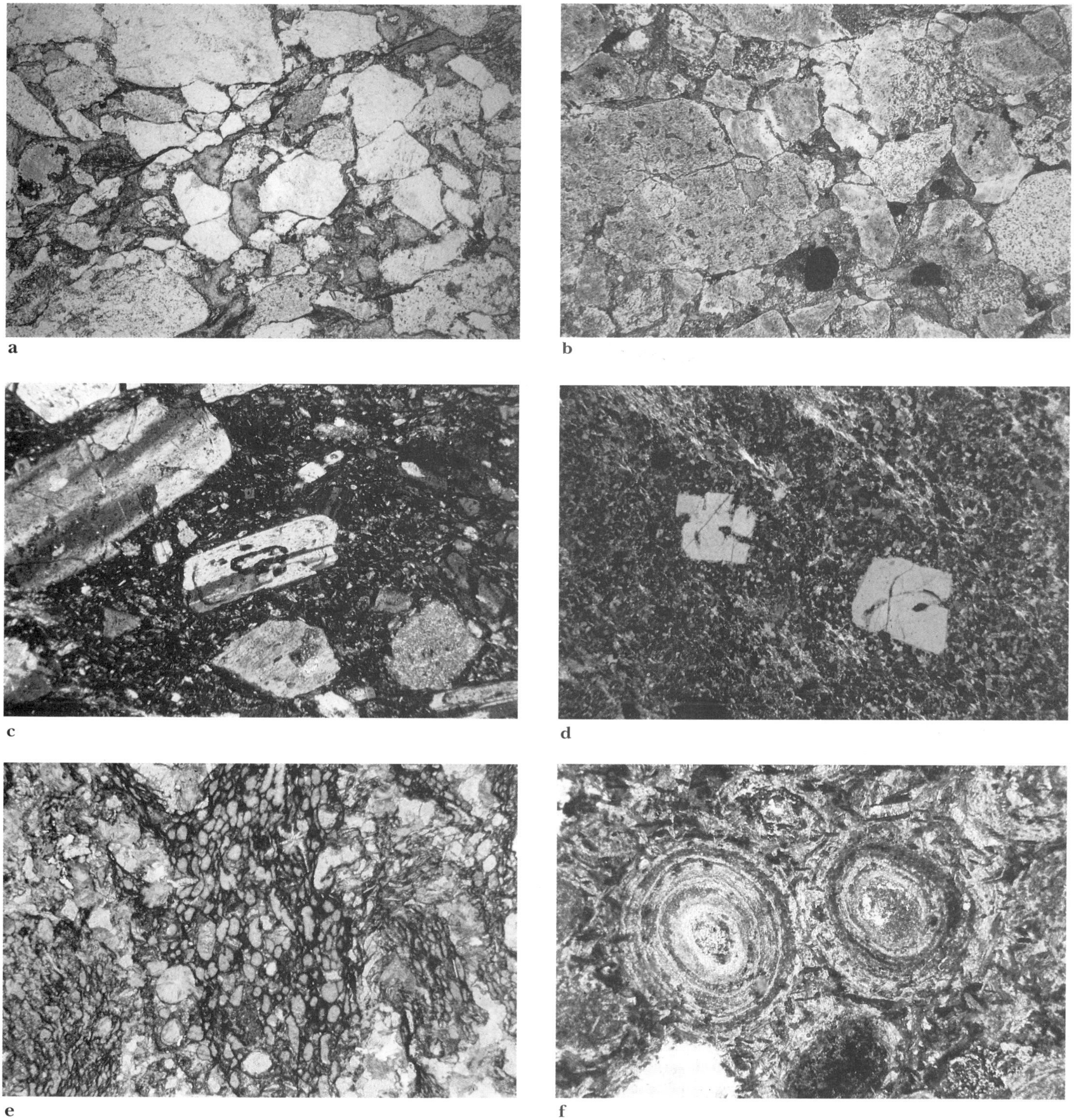

Plate 5 Photomicrographs of Aran Volcanic Group lithologies from formations beneath the Fron Newydd Member.

a) Coarse quartzose sandstone at base of Allt Lŵyd Formation (E65582). Friog [6240 1219]. Field of view about 6 mm wide.
b) Feldspathic sandstone from Allt Lŵyd Formation (E62037). West of Bryn Brith [6551 1505]. Field of view about 4 mm wide.
c) Boulder of feldsparphyric, andesitic basalt from conglomerate in the Allt Lŵyd Formation (E65583). Friog [6227 1185]. Field of view about 6 mm wide.
d) Acid tuff with embayed crystals of quartz. Offrwm Volcanic Formation (E67067). South-west of Castell [5797 0428]. Field of view about 6 mm wide.
e) Ferritised vesicular basalt fragments in basic tuff of the Cefn Hir Member (E62064). Pared y Cefn Hir [6647 1509]. Field of view about 6 mm wide.
f) Oolitic ironstone. Fron Newydd Member (E66201). Ffordd ddu [6476 1281]. Field of view about 4 mm wide.

The mudstones are mainly dark grey or black, locally graptolitic, rusty weathered, well cleaved and bedding is commonly difficult to distinguish. Typical mudstones are well exposed in small quarries and levels, at the type locality, around Llynnau Cregennen [6577 1436 and 6667 1540], and at Penrhyn-gwyn [7028 1493]. In places, the mudstones contain isolated fragments of volcanic rock and feldspar crystals, indicating possible debris-flow deposits. Locally, as in Afon Gwril [6179 0890], interbedded tuffaceous mudstone beds, up to 0.5 m thick, include fragments of acid tuff up to 15 mm across.

West of the Llanegryn Fault, the formation is poorly exposed and consists of mudstones with siltstones, sandstones and conglomerates. Most exposures occur in close proximity to dolerite intrusions and are hornfelsed. Near Bodowen [5892 0649], blue-grey hornfelsed siltstone, 3 m thick, overlain by 5 m of dark grey, cleaved silty mudstone with thin interbeds of laminated siltstone are exposed. Near Tonfanau Beacon [5756 0372], a conglomeratic debris flow deposit, up to 1 m thick, comprises well-rounded acid tuff, tuffaceous sandstone and mudstone pebbles, 5–20 mm in diameter, in a mudstone matrix.

Bryn Brith Member

The member is well exposed at Bryn Brith [6642 1534], the type locality, and consists mainly of massive, lenticular beds of disturbed basic tuff and tuffite. It can be traced from near Arthog to the Cregennen microgranite and for most of this distance it forms a prominent ridge. South of Arthog [6414 1368], three sets of tuffite and tuffaceous sandstone beds, each up to 15 m thick, separated by mudstones, possibly represent infilled channels (Figure 13). The beds wedge out to the west; eastwards [6457 1376] they merge into a single bed, 15 m thick. The massive tuffites are composed of abundant feldspar crystals and angular fragments, 10–50 mm across, with rare boulders, up to 0.5 m, of vesiculated basalt and acid tuff, supported in a mudstone matrix (Plate 4b). In thin section, acid shards, basaltic scoriae, feldspar laths, equant quartz crystals, and rare chamosite ooliths can be discerned in the matrix (E65594). The tuffaceous sandstones are thickly bedded, fine grained and normally graded. Locally, abundant matrix-supported, angular feldspar fragments can be distinguished. The sandstone beds display intense convolute lamination indicating extensive thixotropic disturbance and, elsewhere, slump folding.

North of Llynnau Cregennen [6568 1477], the member comprises about 40 m of coarse-grained, basic tuffs with interbedded mudstones (Figure 13, log A). Towards the north-east the tuff beds merge so that at Bryn Brith, 1 km along strike, the member comprises a massive, coarse-grained, poorly sorted basic tuff, 55 m thick, overlain by 10 m of finer-grained, planar-bedded tuff (Figure 13). The coarse basic tuff contains abundant matrix-supported, subangular fragments, 0.03–0.3 m across, of bedded basic tuff. Contorted and folded bedding indicates that many clasts were unlithified when they were incorporated into the tuff. The matrix comprises ragged, vesicular, ferritised and chloritised basalt fragments, up to 5 mm across, pale green chlorite, opaque oxide and few feldspar crystals (E62052–6). Locally [6605 1501, 6657 1550], the base of the member is marked by beds of basic tuff, up to 1 m thick, with interbedded mudstones. The brown-grey tuffs are composed of subangular to subrounded basalt fragments, up to 15 mm across, some of which are highly vesicular (E62050), and silty mudstone fragments up to 45 mm in length. The tuffs are both normally and reverse graded, and some have planar-bedded tops.

Although correlation is uncertain, a sequence of tuffs and tuffites between Cefn-yr-Owen and Craig y Castell, immediately south and east of the Cregennen microgranite, may be the eastward extension of the Bryn Brith Member. It comprises 50–150 m of massive, blue-grey, coarse-grained, poorly sorted basic tuffs, with few mudstone intercalations. Highly vesicular basalt fragments, up to 10 mm across, are abundant and locally mudstone fragments are common (E62085).

Cefn Hir Member

The Cefn Hir Member comprises basic and acid tuffs and tuffites with interbedded mudstones and rare pillowed basalt. At the type section at Pared y Cefn-hir [6622 1492] (Plate 4c), the sequence is coherent and contains few intervening mudstones, but, farther west, mudstone intercalations are common and the sequence is disrupted by slumping. Locally [6467 1336], the member contains complex basalt and dolerite intrusions/extrusions, which also inhibit detailed internal correlation.

South of Arthog, the base of the member is marked by a massive, basic tuffite, 3–5 m thick, which can be traced for 2 km from Bronllety [6365 1300] to near the Arthog Fault [6450 1370], where it terminates in an area of large-scale, syndepositional disturbance. The tuffite comprises numerous, mudstone matrix-supported fragments of basalt and is overlain by about 10 m of mudstone, and 20–25 m of massive, locally welded, acid tuff. These tuffs contain euhedral, and fragmented, quartz and feldspar crystals, and, in places, many recrystallised acid volcanic rock fragments (E65603, E66352). In places, the acid tuffs form isolated, lenticular beds and pods [6365 1285]. At the top of the member, basic tuffs are interbedded with the mudstone; the tuffs are generally up to 2 m thick, but locally, as near Bron-lletty [6372 1295], they are up to 9 m thick.

At the south-west end of Pared y Cefn-hir, the member, 42 m thick, consists of basic tuff beds, mainly 1–15 m thick, with interbedded blue-grey tuffaceous mudstone 2–3.5 m thick (Figure 13). An acid ash-flow tuff, 2 m thick, can be distinguished in the middle of the section and a pillowed basalt, approximately 5 m thick, occurs near the top. The distinctively rusty weathered basalt is clearly featured on the south-west end of the ridge. To the north-east, the lower basic tuffs wedge out [6653 1520] and the base of the member is marked by the thin acid tuff. This displays a crystal-rich basal zone, with feldspar (50 per cent) and quartz (5 per cent) crystals, and acid volcanic rock fragments up to 10 mm in diameter, which grades up into finer-grained tuff with fewer feldspar crystals (10 per cent) and a weakly developed welding foliation (E66789–90). The uppermost 25 m of the member consists of massive, channelised, poorly sorted and block-rich

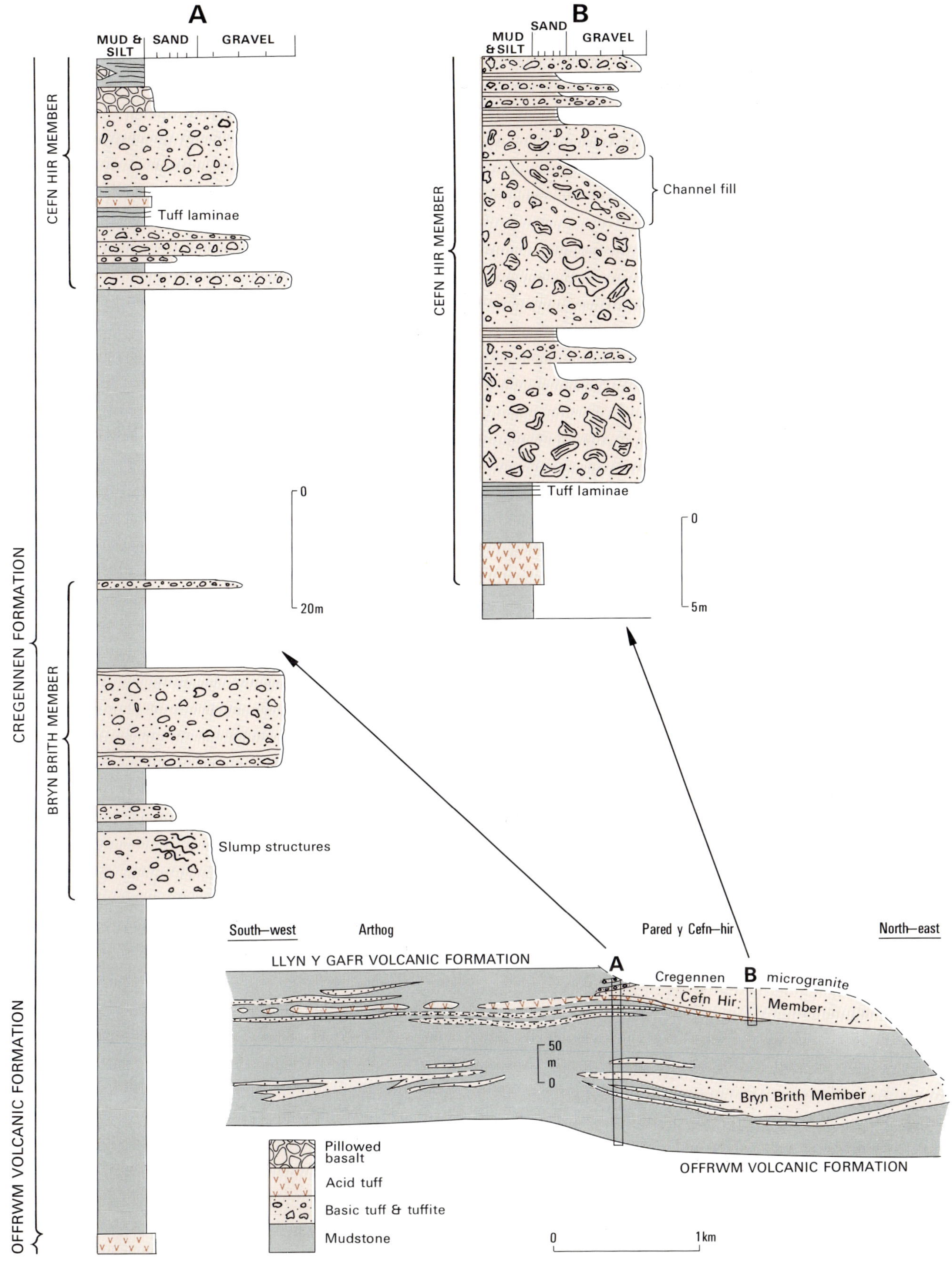

Figure 13 Generalised lithological variations within the Cregennen Formation between Arthog and Pared y Cefn-hir. Log a) is from Gefnir Farm and Pared y Cefn-hir [6572 1473 to 6610 1475]. Log b) is from the east end of Pared y Cefn-hir [6653 1520].

basic tuffs with intercalated planar-bedded, basic tuffs (Figure 13; Plate 6). Individual beds are commonly, normally graded close to their tops. Angular to subangular blocks, up to 0.5 m across, commonly tabular, are mainly of basalt, bedded basic tuff, silty mudstone and acid volcanic rock. Whole and fragmented crystals of feldspar and resorbed quartz (E66793) are locally common.

Near Nant-y-gwyrddail, south of the Cregennen microgranite, the possible continuation of the member comprises about 100 m of basic tuff and tuffite with a significant proportion of interbedded tuffaceous mudstone. The massive, normally graded, poorly sorted, basic tuffs are up to 2.5 m thick [6744 1487], and are composed mainly of subangular to rounded clasts, up to 50 mm across, of basalt, acid volcanic rock, quartzose sandstone and scattered feldspar and quartz crystals (E61110–1).

East of the Ceunant Fault, around Mynydd-y-gader, basic and acid tuffs crop out and are considered to be the lateral equivalents of the Bryn Brith and Cefn Hir members. Direct correlation is inhibited by complex mudstone interdigitations, numerous, irregular igneous intrusions and large-scale slumping, as around Tyddyn-mawr. However, on the south side of Mynydd-y-gader [7253 1487] the sequence is less disturbed and acid tuffs of the Offrwm Vol-

Plate 6 Bedded, basic tuffs overlain by coarse, basaltic debris flow deposit. Cefn Hir Member. Pared y Cefn-hir [6652 1516].

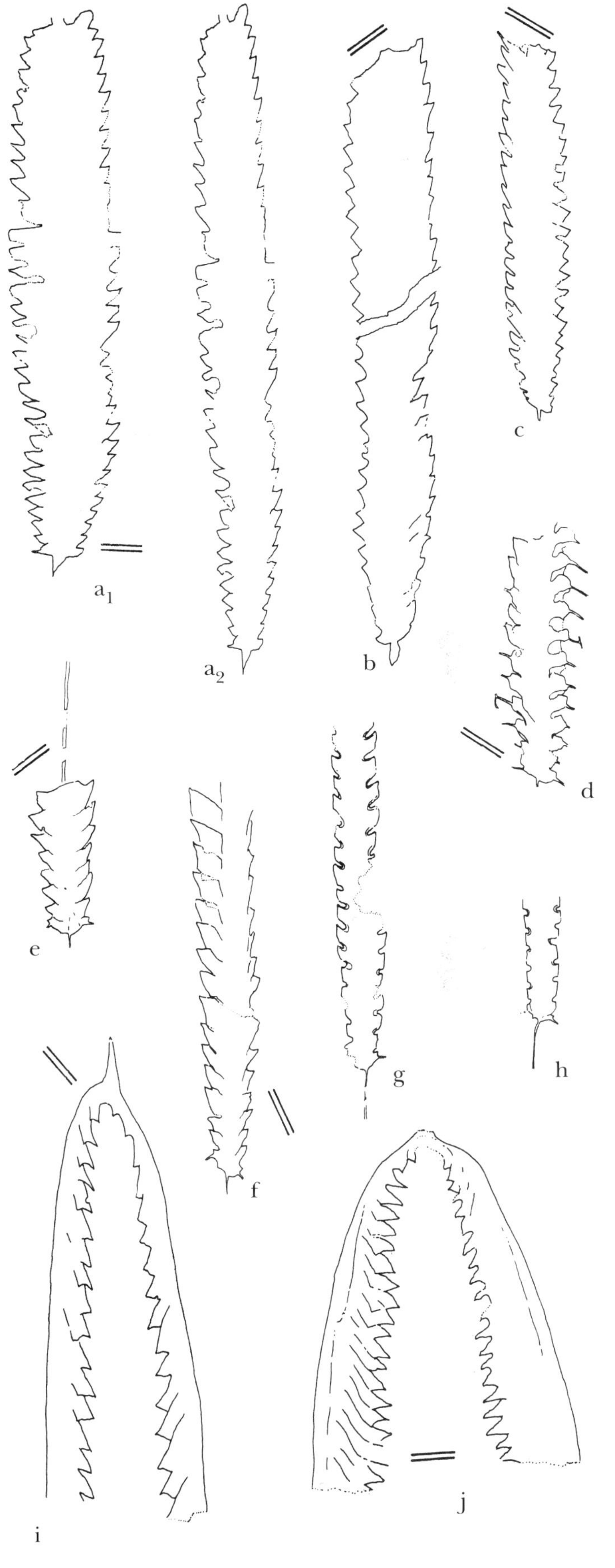

Figure 14 Graptolites from the Aran Volcanic Group. All the specimens are preserved in low relief and are magnified × 5. The parallel lines indicate the trace of cleavage on the plane of bedding.

a), b), c) *Diplograptus foliaceus* (Murchison), all on block RX 3969. 1), compressed, 2) with deformation removed. Benglog Formation, Caradoc (*multidens* Biozone), east of Cross Foxes [7888 1634].
d) *Lasiograptus costatus* Lapworth?, RX 4399, partly restored from counterpart. Ty'r Gawen Mudstone Formation, Caradoc (*multidens* Biozone), east of Llyn y Gadair [7109 1374}.
e), f) *'Glyptograptus' teretiusculus* (Hisinger) of Elles & Wood, RX 4331c, 4332. Ty'r Gawen Mudstone Formation, Caradoc (*multidens* Biozone), east of Llyn y Gadair [7109 1374].
g), h) *Amplexograptus*? *molestus* (Thorslund), National Museum of Wales 79.50G.27a (A H Cox collection). Ty'r Gawen Mudstone Formation, Caradoc (*multidens* Biozone), west side of Tyrrau Mawr [c. 6732 1332].
i), j) *Didymograptus (D.) spinulosus* Perner? RX 3235, 2327. Cregennen Formation, Llanvirn (*artus* Biozone), slate trial north-east of Bryn Brith [6668 1542].

canic Formation are overlain by 60 m of massive, coarse-grained, poorly sorted, basic tuffs with mainly subangular to subrounded basalt and acid volcanic rock fragments, 4–150 mm across. Angular clasts of slumped, laminated siltstone, up to 0.5 m across, occur near the base [7233 1479] (Plate 4b). Some interbedded, laminated basic tuffs are apparent in the upper 10–15 m. This sequence is similar to the Bryn Brith Member and is overlain by about 100 m of mudstones with impersistent intercalations of coarse-grained, basic tuff. Near Penrhyn-gwyn slate quarries [7047 1488], where the succession is more complex, the mudstone intercalation is thicker and includes a massive, weakly foliated acid tuff, about 25 m thick, probably equivalent to those of the Cefn Hir Member. In thin section, bimodal shards are well defined and fragments, of welded tuff and pumice up to 6 mm across, occur within a fine-grained, quartzose aggregate after vitric dust (E62973–4).

BIOSTRATIGRAPHY

The mudstones of the Cregennen Formation have yielded Llanvirn fossils at numerous localities. Trilobites obtained from mudstones in the Bryn Brith Member, southwest of Bryn Brith [6581 1486], include *Stapeleyella murchisonii* (Plate 9h, l), with fragments of *Barrandia*?, *Degamella*?, *Ectillaenus*? and an eccoptochiline? *S. murchisonii* suggests an early Llanvirn age.

Graptolites were collected from the mudstones between the Bryn Brith and Cefn Hir members at several places. *Didymograptus* (*D.*) cf. *spinulosus* (Figure 14i, j) was found west of Llynnau Cregennen [6578 1437] and at a slate trial east of Bryn Brith [6668 1542], and *D.* (*D.*) *stabilis* from south of Bryn Brith [6625 1506]. South of Kings [6844 1543] and in the Gwynant valley [6884 1548], poorly preserved pendent Didymograptids are associated with trilobite fragments. To the north-east, *D.* (*D.*) *artus* was collected south of Tabor [7512 1674]. All the forms of *Didymograptus* identified are suggestive of the *artus* Biozone (Table 2).

South of Arthog [6467 1334], graptolites from the mudstone between the Cefn Hir Member and the base of the overlying Llyn y Gafr Volcanic Formation include *'Amplexograptus' confertus*?, and fragments of *Climacograptus*?, *Cryptograptus*?, *Didymograptus* (*D.*) sp., *'Glyptograptus'* and *Pseudoclimacograptus*?. These are of Llanvirn age but of uncertain zone. Cox collected a pendent *Didymograptus* together with a dalmanellid? brachiopod from about the same horizon south of the intrusion on Mynydd-y-gader [about 720 145].

CONDITIONS OF DEPOSITION

Most of the basic tuffs and tuffites are probably debris-flow deposits. The amount of basaltic debris indicates contemporaneous explosive basaltic activity although its source is not easily recognised. However, the rare pillowed lavas and thin basalt intrusions suggest the district lay close to the volcanic centre. The north-eastward thickening of the tuffs, and wedging out of the intervening mudstones, suggests it lay in this direction, possibly coincident with the postulated uplifted area which affected deposition of the Allt Lŵyd and Offrwm Volcanic formations.

The coarser-grained tuffites probably represent broad channel-fill deposits and the arrangement of lenticular beds, as south of Arthog (Figure 13), suggests that channel axes probably migrated laterally. The numerous tabular and contorted clasts of bedded tuff incorporated within the debris-flow deposits suggest erosion of channel margins and incorporation of the substrate. The upward-fining and upward-thinning, tuff sequences on Pared y Cefn-hir represent the progressive filling of the channels. The thinner, finer-grained, muddy tuffites are more laterally persistent and indicate the development of a relatively smooth, gently inclined depositional surface following the channel infilling phase. The thinner-bedded and graded basic tuffs are interpreted as turbidites. Disrupted and slipped rafts of strata, such as those south of Arthog, indicate large-scale sliding of beds probably caused by contemporaneous faulting.

Brithion Formation

The Brithion Formation comprises up to 110 m of acid tuff and conglomerate. It crops out only in the area north of Cross Foxes, at the edge of the Harlech district, where it was first defined (Allen and Jackson, 1985), and wedges out south-west of Afon Clywedog.

East of Caerynwch the formation, 30–110 m thick, consists of an acid ash-flow tuff, 30 m thick [7715 1767], overlain by poorly exposed conglomerate, 0–80 m thick [7706 1741 to 7726 1763]. The tuff is massive with euhedral and fragmented crystals of ablite-oligoclase and quartz in a devitrified and recrystallised quartzose, shard-rich, matrix (E65140). The conglomerate is also massive and composed of rounded clasts of massive and banded acid tuff and/or rhyolite, 0.02–0.35 m in diameter, in a fine-grained matrix of quartzose recrystallised shards, acid volcanic rock fragments, and chlorite [7719 1756] (E65143, E65144).

Along Afon Clywedog [7660 1723], at the southern edge of the outcrop, the acid ash-flow tuff, up to 10 m thick, conformably overlies mudstones of the Cregennen Formation. The tuff contains fragments of feldsparphyric rhyolite, with a flow-foliated groundmass (E65912), and perlitic fractured rhyolite (E65913, E65915) and is conformably overlain [7661 1722] by a tuffaceous mudstone, debris-flow deposit, 2–3 m thick, with acid tuff clasts up to 0.25 m in diameter.

CONDITIONS OF DEPOSITION

The acid ash-flow tuff is considered to have been derived from an extrusive rhyolite dome at Creigau Brithion in the Bala district (Dunkley, 1978, 1979; British Geological Survey, 1986). Its apparent conformity with underlying and overlying mudstones suggests that its emplacement was subaqueous. From the lithologies, lack of bedforms and thickness variation, the conglomerate is interpreted as the infill of a channel.

Llyn y Gafr Volcanic Formation

The Llyn y Gafr Volcanic Formation comprises mainly basalt lavas and basic tuffs with few acid tuffs and thin mudstones. The basalt lavas represent the first major effusive volcanic episode within the district. It is broadly equivalent to the Lower Basic Volcanic Series and Llyn-y-Gafr Spilitic Group (Cox and Wells, 1921, 1927; Cox, 1925) and the Basic Volcanic Group of Jones (1933).

The formation is thickest, up to 360 m, between Tyrrau Mawr and Mynydd Moel, but it thins to the south-west and it is absent west of the Llanegryn Fault, possibly removed by erosion prior to deposition of the Fron Newydd Member (Figure 7). Accurate thickness estimates are inhibited by numerous high-level, basalt/dolerite intrusions, which are commonly difficult to distinguish from extrusive basalts. Internal correlation of the lavas and tuffs, including acid ash-flow tuffs, is imprecise because of marked lateral facies variation, but good sequences can be established locally, as south of Arthog, at Llyn Gafr (the type section [7110 1300]) and north of Mynydd Moel.

The main lithologies are massive and pillowed, aphyric basalts, locally with reddened, oxidised tops, and intercalated tuffs, but in the east the formation becomes dominated by tuffs. Many tuffs are an admixture of acid and basic pyroclastic debris.

In the west of the district, exposure is poor but the outcrop is locally well featured [6250 1000]. East of the Llanegryn Fault [6100 0680], the formation comprises 150–200 m of basalts with interbedded basic tuffs and mudstones, but the thickness is probably exaggerated by basic intrusions. Near Nant Gwril [6198 0842], about 3 km to the north-east, in the middle of the outcrop are 15 m of massive, fine-grained, planar laminated, basic tuffs with scattered, elongate clasts, up to 50 mm long, of vesiculated basalt concentrated in layers. Nearby [6261 0946], 3 m of planar laminated basic tuff overlies pillowed basalt at the top of the formation.

South of Arthog the formation is 160 m thick and the lowest 45 m comprises alternating beds, up to 2 m thick, of massive, basic tuff and massive, non-welded, acid tuff with basalt and acid tuff clasts and rare iron oxide ooliths

(E66196). Above, two basalts are separated by a sequence, up to 20 m thick, of massive tuffite and planar laminated basic tuff. The lower basalt, 65 m thick, probably comprises several flows, which are massive, locally brecciated and progressively more jointed and rusty weathered towards the top. The upper basalt, 30 m thick, is pillowed throughout and is overlain by 2 m of tuffite and planar laminated and cross-stratified basic tuff.

South of Llynnau Cregennen, the sequence is intruded by a complex of dolerite sillls, but locally, as near Hafottyfach [6615 1382], a pillowed basalt, up to 50 m thick, can be distinguished. It is overlain by a bleached, massive acid tuff, 15 m thick, with a well defined, bedding-parallel welding foliation. The tuff contains quartz and feldspar phenocrysts in a devitrified and recrystallised quartzose aggregate with a well-developed eutaxitic fabric (E61101). Also in this vicinity [6653 1497], pillowed basalt forms a raft, about 200 m long, within the Cregennen microgranite.

On Tyrrau Mawr [6815 1411], the formation is about 300 m thick, but the lower part is poorly exposed and intruded by a microgranite. The upper part is a massive basalt breccia, up to 140 m thick, which thins to 75 m at Llyn Gafr, approximately 3 km to the east (Figure 15). The breccia comprises poorly sorted, clast-supported, angular blocks, generally 5–200 mm across, of highly vesicular basalt with a sparse matrix of basic tuff. The lowest 20 m of breccia apparently grades from the underlying pillowed basalts [6810 1432] and is crowded with blocks, up 1 m across, of massive and pillowed basalt.

Around Llyn Gafr, basalts up to 70 m thick, probably composite sequences of several lava flows, are intercalated with basic tuffs and mudstone (Figure 15). Lavas near the base are generally massive with pillowed tops and those in the upper part are mainly vesicular and pillowed with massive basal layers 5–10 m thick. Pillows are most commonly 0.5–0.75 m across, rarely [7321 1431] up to 3 m, and are closely packed, with little or no interpillow sediment. The cores are crowded with vesicles and carbonate amygdales; chilled margins are mainly 10–20 mm thick. The basic tuffs are coarse grained, poorly sorted, with basalt clasts 2–150 mm across. They form sequences, 2–25 m thick, of mainly massive beds, but some are normally graded, few are reverse graded and in places alternating coarse- and fine-grained beds, up to 0.25 m thick, can be distinguished (Figure 15).

North of Mynydd Moel, the formation is about 290 m thick and its base is marked locally [7213 1473, 7336 1480, 7422 1536] by a thin, impersistent basalt with numerous feldspar phenocrysts up to 5 mm long (E65955) (Figure 15). Basic tuffs within the overlying lavas comprise alternating layers, 5–30 mm thick, of vesicular basalt fragments, 2–3 mm across, and finer shards (E66797). The beds are mainly massive and planar laminated but some show large-scale, low-angle cross-stratification.

In the vicinity of Bwlch-coch [7492 1586], the basalt lavas wedge out and the formation, 225 m thick, is mainly of basic and acid tuffs. A basal basic tuff sequence, up to 50 m thick, comprises coarse-grained beds of angular basalt fragments, 5–30 mm across, with clasts, up to 60 mm, of partly carbonated feldsparphyric basalt and mudstone [7463 1565, 7482 1594] (E65199). Above, a massive, non-welded, acid tuff, 10–20 m thick, comprises albite crystals, 1–2 mm long, and ragged pumice fragments up to 25 mm across, in a shard-rich matrix [7501 1604] (E65179). Mudstones, about 15 m thick, above the acid tuff are in turn overlain by planar-laminated and fine-grained basic tuffs, 90 m thick. The top of the formation is marked [7520 1594] by a massive acid tuff, up to 40 m thick, which can be traced for 1.6 km along strike. It is similar to the lower acid tuff, but also contains sub-angular acid and basic volcanic rock fragments, up to 50 mm across (E65158, E65187) and is locally [7523 1596] rich in feldsparphyric basalt fragments (E65156).

A short distance from Bwlch-coch, near Cross Foxes, the formation thins to less than 50 m and is poorly exposed. Along Afon Clywedog [7663 1719], 20 m of grey-green, fine- and coarse-grained, carbonated basic tuff, in beds 4 mm to 0.5 m thick, display ripple-marked bedding planes. The tuffs are deeply altered although shards are locally distinguishable (E65916).

East of the Bala fault, between Tyddyn Du [7663 1518] and Cae'r-tyddyn [7890 1738], the formation, 330 m thick, consists mainly of basic tuff with some interbedded mudstones but it is poorly exposed. South of Cae'r-tyddyn, green-grey, basic tuffs comprise subangular to rounded clasts of feldsparphyric basalt, 2–90 mm in diameter (E65036, E65113, E65116), and a few of mudstone, rhyolite [7703 1556, 7795 1636] and acid tuff (E65036) also occur. Typical of the upper part of the formation is a well-exposed sequence at Bwlch Oerddrws, about 100 m thick, of alternating massive tuff, graded tuff beds, and sequences of thinly interbedded, acid tuff and mudstone (Figure 16). The thinly bedded sequences display numerous small-scale, pre-lithification faults and slump structures.

Petrography

The massive lavas are petrographically similar to the basalt intrusions, with altered augite crystals, up to 5 mm across, feldspar laths, less than 1 mm in length, and much interstitial chlorite, epidote and leucoxene/sphene. The augite crystals, with intense actinolite alteration, partly enclose feldspar laths in a subophitic texture (E65185). The original calcic feldspars are albitised and sieved with inclusions of calcite, chlorite and epidote. Pillowed and vesicular basalts with flow-foliated feldspar microlites, locally in radiating sheafs (E65223–5), are generally finer grained than the massive basalts. The chilled margins of pillows consist of chloritised basaltic glass (E65224).

The fine-grained basic tuffs, and the matrix of the coarser-grained varieties, comprise recrystallised, chloritised and ferritised, cuspate shards and scoriae of basaltic glass, up to 1 mm long, in a chloritic matrix (E66197).

Biostratigraphy

The only fossils collected from the Llyn y Gafr Volcanic Formation are from mudstones, 1 m thick, exposed alongside the Pony Path 1.2 km north-west of Llyn y Gadair [6954 1418]. *Cryptograptus*, *Dictyonema*? and sever-

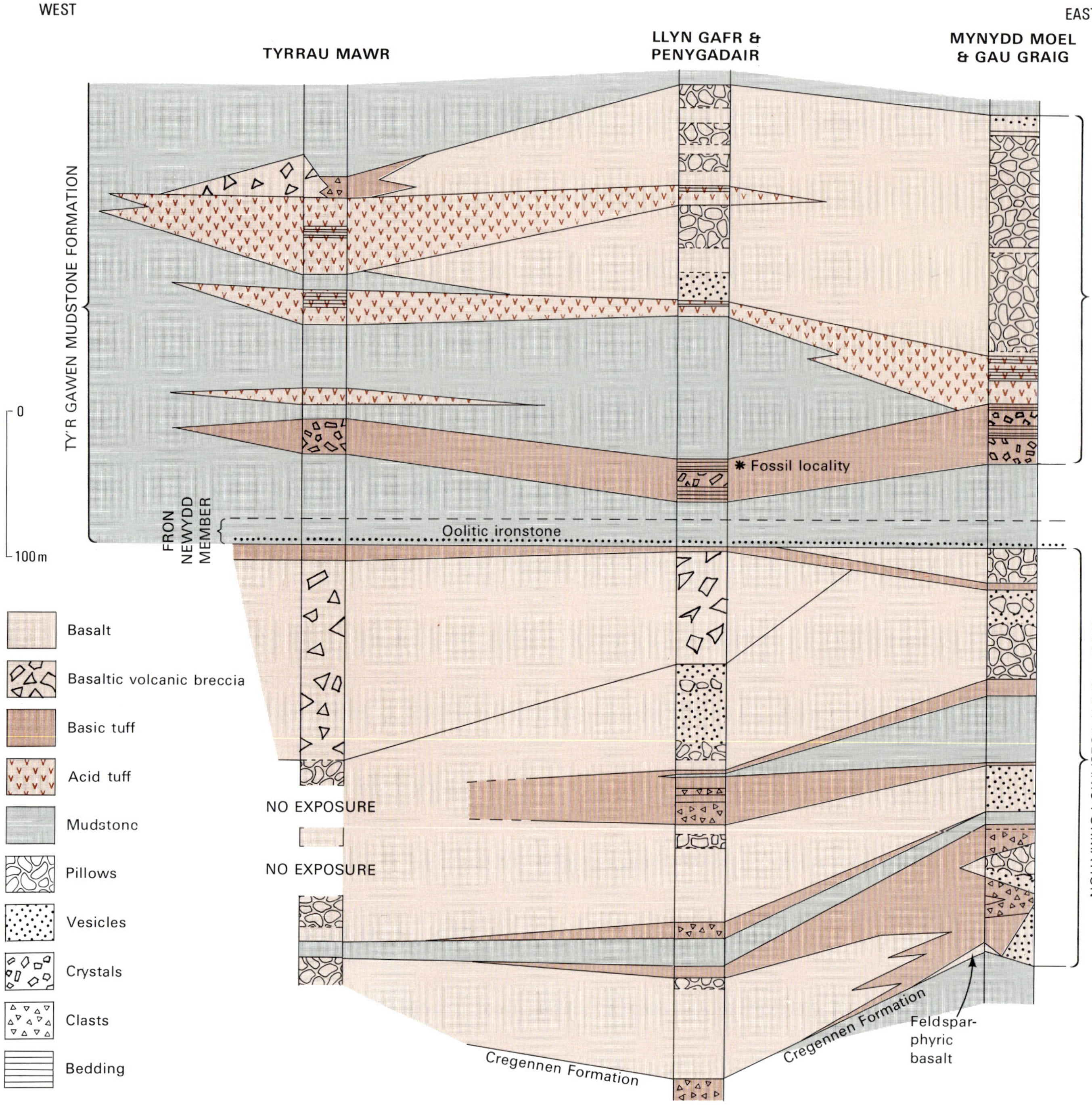

Figure 15 Generalised vertical sections of the Llyn y Gafr and Pen y gadair Volcanic formations in the central part of the outcrop. Tyrrau Mawr [6745 1348]. Llyn Gafr and Penygadair [7090 1275]. Gau Graig [7252 1497 to 7403 1445] (see Figure 7 for location of sections).

al diplograptids were obtained, but all are badly preserved and their age is uncertain.

Conditions of deposition

The massive basalt lavas, which dominate the formation, reflect gentle effusion. The pillowed basalts and intercalated graptolitic mudstones indicate a marine setting and the paucity of evidence of in-situ reworking suggest that this was well below wave base. The wedging out of basalts around Cross Foxes suggests that they were ponded to the west, probably by syndepositional faults. The abundance of scoriae and cuspate fragments, and the frequency of the basic tuffs, suggests that the pyroclastic debris was not the result of chilling and decrepitation of the basalt lavas during transport, but resulted from explosive eruption. However, the bed forms suggest that pyroclasts were subsequently reworked by debris flows and turbidity currents, and probably infilled depressions

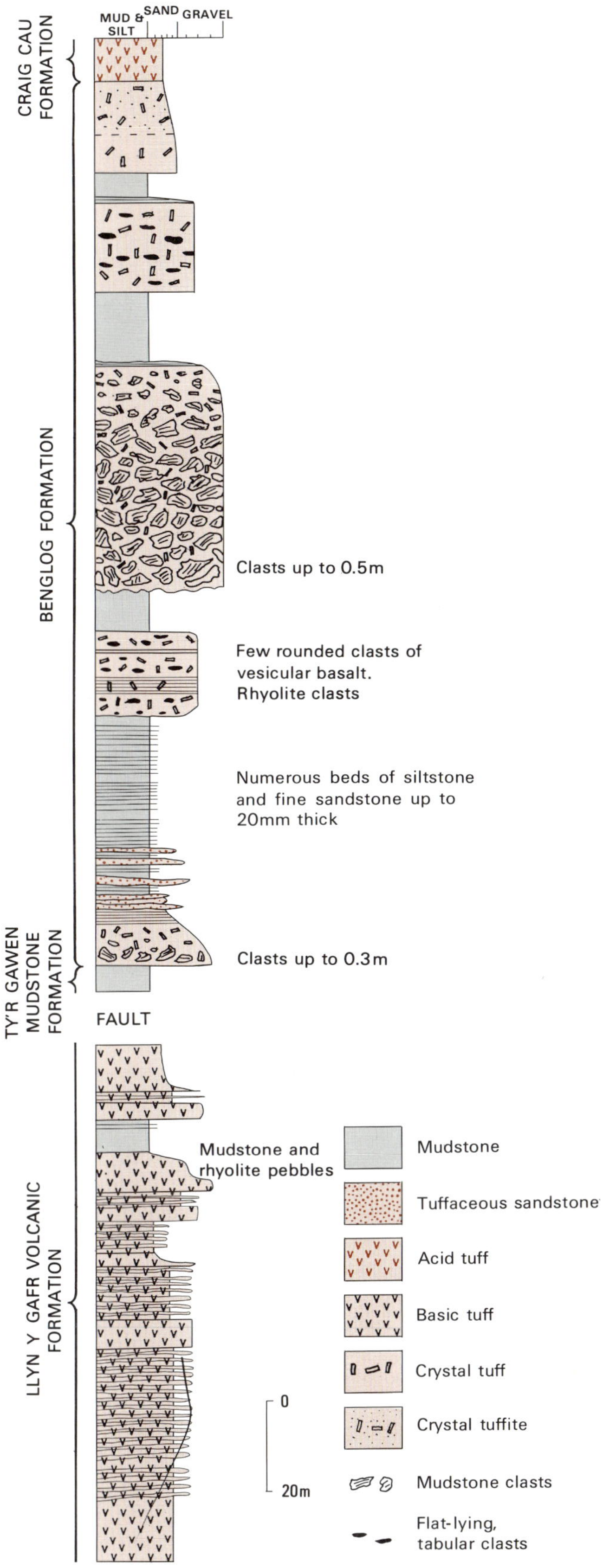

Figure 16 Log of the Llyn y Gafr Volcanic and Benglog formations around Bwlch Oerddrws [7920 1685 to 7972 1700].

between basalt flows. The locally thick sequences of basic tuffs which are planar laminated throughout, or display large-scale, low-angle cross-stratification, probably reflect gradual accumulation of coarse and fine pyroclasts by settling and/or transport by dilute turbidity currents onto the irregular upper surfaces of basalt flows.

The hyaloclastites probably formed both by spalling of glassy selvages of the pillows and by fragmentation of basalt pyroclasts as a result of rapid chilling in contact with sea water (Fisher and Schmincke, 1984; Kokelaar, 1986). The thick sequence of basaltic, pyroclastic breccia near Llyn Gafr is unlike an autobrecciated, 'aa'-type basalt flow and possibly represents a block accumulation close to a vent.

Ty'r Gawen Mudstone Formation

The Ty'r Gawen Mudstone Formation, previously the Llyn y Gader and Llyn Cau mudstones (Cox, 1925; Cox and Wells, 1927), consists of up to 1 km of mudstones with few siltstones and sandstones, some of which are tuffaceous. Locally, lateral disruption of the sequence resulted in the development of olistostromes. East of the Llanegryn Fault, the base is conformable and marked by the Fron Newydd Member, which comprises phosphatic mudstones and oolitic ironstone, but west of the Llanegryn Fault the base is an angular unconformity. In the east and central part of the outcrop, the formation interdigitates with the Pen y gadair Volcanic Formation (Figures 7, 15).

Fron Newydd Member

Over most of the outcrop, the Fron Newydd Member comprises about 25 m of black, massive, pyritic mudstone with numerous phosphate nodules and scattered ooliths. In many places, it conformably overlies the Llyn y Gafr Volcanic Formation and contains an impersistent oolitic and pisolitic ironstone, 1–2 m thick, between 2 and 4 m above the base. Because the ironstone has been worked as a source of low-grade iron ore, the outcrop is marked by pits, adits and small quarries. Good exposures occur in pits at Ffordd Ddu [6476 1283], between Llyn y Gadair and Mynydd Moel [7099 1375, 7476 1547, 7510 1569] and at the type locality near Llanegryn [6055 0598 to 6058 0583]. The thickest development of ironstone, 6–10 m, is poorly exposed in overgrown quarries near Cross Foxes [7598 1640, 7608 1655] and south-east of Bwlch-coch [7496 1556] (Strahan et al., 1920), but the member has not been distinguished east of Cross Foxes.

The ironstone is mainly a massive, structureless bed of grain-supported chamosite ooliths, 0.5–1 mm in diameter, with rare pisoliths up to 20 mm in length. It is possible that several thin, impersistent ironstone beds are developed. The ooliths are generally concentrically banded, with rare cores of ferritised basalt or angular quartz, and are locally overprinted by radial sprays of acicular carbonate (E66201) (Plate 5f). In many ooliths, the chamosite has oxidised to limonite. Ferritised faecal pellets, less than 0.1 mm in diameter, are common and are locally the dominant component (E67052).

In the south-west of the district, near the Gwril valley, the ironstone wedges out and the pyritic mudstones thicken markedly. The mudstones are rusty weathered, iridescent and poorly cleaved, with scattered, irregular and ellipsoidal, phosphate nodules, up to 80 mm long. Nodule shapes vary from amoeboid to oblate ellipsoids and their surfaces are commonly shiny and grooved. The mudstones also contain angular quartz grains and much disseminated pyrite. Chamosite ooliths, very similar to those within the oolitic ironstone, occur mainly within the phosphate nodules (E67050, E67058, E67024, E67028), but also form rare, grain-supported lenses, up to 5 mm thick. The outcrop is distinctively ridged, possibly reflecting the alternation of silt, pyrite or phosphate nodule content in the mudstones.

Near Peniarth [6050 0600], immediately east of the Llanegryn Fault, the member is 170 m thick and is conformable on the Llyn y Gafr Volcanic Formation. However, west of the fault the base is unconformable. The unconformity is greatest at Llwyngwril, where the member, about 180 m thick, overlies the Allt Lŵyd Formation (Figure 7). In comparison with the sequence on the east side of the fault, 300–400 m of strata are missing, but to the south-west, near Rhoslefain, this has diminished and the member overlies the Cregennen Formation. In this vicinity, thin beds and lenses of sandstone and tuffite within the mudstone indicate slump disruption. Angular grains of sodic plagioclase, quartz and altered basaltic glass are the major components (E67053, E67023). Near Bodowen [5884 0620], black mudstones with phosphate nodules include a 2 m-thick bed of friable tuffaceous sandstone, with rip-up clasts of mudstone and abundant acid shards (E67026). Near Tonfanau the member increases in thickness to about 55 m.

Strata above the Fron Newydd Member

These are mainly dark grey, rusty-stained mudstones with rare sandstones and tuffaceous rocks. In much of the formation, bedding is ill defined, although locally there is much evidence of slumping. The upper part includes thin, well-bedded sandstones. Exposure is generally poor, but is good around the Ty'r Gawen type section [6240 0660], the Cadair Idris escarpment, Craig Cau [7100 1245], Craig Cwm-llwyd [6425 1200] and Craig-y-llyn [6650 1200].

In the west of the district the sequence is thickest, about 900 m, with little volcanic debris, but to the east it thins and is split into upper and lower parts by the Pen y gadair Volcanic Formation (Figure 7). Near Graig Gau [7450 1430], 3 km east of Cadair Idris, the lower mudstone sequence wedges out and the upper sequence is reduced to 35 m. East of the Bala Fault, only 60 m of mudstones, locally with thin sandstones, as south of Nant Ffridd-fawr [7761 1586], occur between the Llyn y Gafr Volcanic and Benglog formations (Figure 7).

Sandstones are rare in the lower part of the formation and are only developed near Tonfanau [5812 0394], where, 50 m above the Fron Newydd Member, a 3 m-thick, massive, coarse-grained, feldspathic sandstone with mudstone rip-up clasts displays well developed bifurcating grazing trails on its base. The sandstone is poorly sorted, with little matrix, and lithic fragments include granophyre, basalt and acid tuff (E67082). The rock is mostly cemented by euhedral quartz overgrowths (cf. Waugh, 1970) and chlorite.

In the western part of the outcrop, sequences of turbiditic sandstone, most less than 10 m thick, occur in the upper 150 m of the formation and can be traced from near Bird Rock [6332 0792] to the north-east side of the Broad Water, near Tywyn. Some are tuffaceous, but most are dominated by quartzose grains. At Gwyddfriniau [5928 0434], west of the Llanegryn Fault, a 7.5 m-thick sequence comprises thickly bedded, normally graded, locally pebbly sandstones, which thin and fine upwards (Figure 17). The thick beds are generally normally graded, with local parallel-lamination towards their tops, whereas the thinner beds are mainly ripple cross-stratified and planar laminated. The subrounded pebbles, up to 60 mm in diameter, are mainly of acid tuff, mudstone, basalt and dolerite with rare quartz-mica schist, quartz-phyric rhyolite and vein quartz (E67047). Flute casts, grooves and ripple cross-stratification mainly indicate derivation from the north, north-west and west (Figure 18).

Near Bryncrug [6153 0393], east of the Llanegryn Fault, two sandstone sequences, 6–7 m thick, are separated by 45 m of mudstone. The upper sequence thins and fines upwards and is probably equivalent to that at Gwyddfriniau. Texturally, the sandstones are poorly sorted, clast-supported greywackes, with over 15 per cent matrix. The grains are mostly angular quartz, albite, alkali feldspar and chlorite-mica stacks. Other detrital components include muscovite, phosphate nodules, tourmaline, epidote, zircon, siltstone, mudstone, acid tuff, basalt and schist (E67060–62).

Near Craig Ty'n y cornel [6436 0866, 6326 0790], 5 m of medium to thickly bedded, basic tuffs and tuffites, occur approximately 150 m below the top of the formation. Similar beds are exposed west of Llyn Cau [7070 1236], within mudstones underlying the Craig Cau Formation; the sequence, 11 m thick, comprises massively to thickly bedded tuffs, tuffites and tuffaceous sandstones. The beds, up to 2 m thick, are mainly normally graded and planar laminated. The tuffs, which contain abundant fragments of chloritised basic glass, altered basalt, acid tuff and rhyolite (E62622–3), wedge out along strike and possibly represent a channel infill. East of Llyn Cau [7196 1239], the equivalent sequence comprises massively bedded, feldspar-crystal-rich, tuffites.

Between Llyn Cau and Graig Gau, there is evidence of extensive lateral disruption in the mudstones above the Pen y gadair Volcanic Formation. In the steep slopes between Craig Cau and Llyn Cau [7098 1258 to 7111 1250], an olistostrome, 200 m long and 20 m thick, occurs 10–20 m above the Pen y gadair Volcanic Formation (Figure 7); it comprises blocks of pillowed basalt, up to at least 4 m across, in a tuffaceous mudstone matrix with variable amounts of volcanic rock fragments and feldspar crystals. The matrix infills cracks in the blocks. Bedding is discernible only in a few places and it is invariably slump folded. Irregular, contorted clasts of pale green tuff, up to 0.3 m across, indicate incorporation whilst unlithified. Along strike, bedding in the mudstones is also disaggregated and disturbed. Near Gau Graig

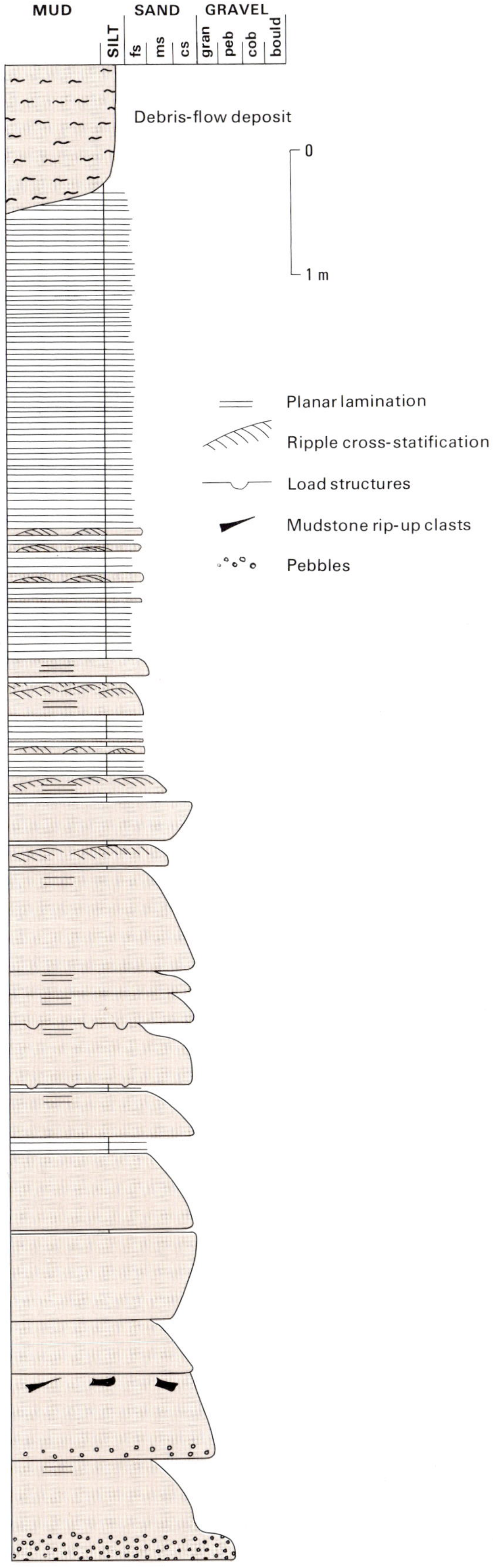

Figure 17 Sedimentary log of a sandstone sequence within the Ty'r Gawen Mudstone Formation. Gwyddfriniau [5928 0434].

[7462 1437], the mudstones include slabs, up to 5 m across, of contorted tuff and tuffite. To the north-east, a complex mixture of mudstone and large basalt blocks, 5–10 m long, can also be distinguished [7478 1462 to 7494 1473].

East of the Bala Fault, between Tyddyn Du and Bwlch Oerddrws [7757 1571 to 7786 1590, and 7859 1637 to 7907 1668], the mudstones include an acid ash-flow tuff, up to 60 m thick, which may be the lateral equivalent of the tuff at the top of the Llyn y Gafr Volcanic Formation at Bwlch-coch. The tuff is massive, with chloritised pumice fragments, up to 15 mm in length. A well-developed, bedding-parallel welding foliation can be discerned on weathered surfaces. In thin section, altered sodic plagioclase crystals within a eutaxitic, shardic fabric (E65033, E65038, E65106) can be distinguished.

In places, mudstones at the top of the formation were disrupted by emplacement of the overlying ash-flow tuffs of the Craig Cau Formation. For example, silicified and noncleaved mudstones at Bird Rock [6450 0670] and Craig Ty'n-y-cornel [6450 0845] display disaggregated bedding and rare slump folds. In thin section, microfolded bedding, a possible compactional alignment of chlorite mica stacks and other inequant grains, and a second grain alignment representing the regional cleavage (S_1) can be determined (E67043, E67039–40). Diagenetic monazite nodules up to 0.2 mm in length are also common.

Biostratigraphy

At some localities the Fron Newydd Member has yielded numerous graptolites, but their poor preservation inhibits precise biostratigraphical interpretation. At a small adit north of Mynydd Moel [7280 1415] the fauna includes fragments of *Dicellograptus*, *'Glyptograptus'* sp. and *Pseudoclimacograptus scharenbergi*?, together with horny brachiopods and spicules. A small excavation south of Bwlch-coch [7446 1513] yielded *Paterula*, spicules and scolecodonts, and debris from workings south-east of Bwlch-coch [7496 1556] yielded *Paterula*? and the graptolites '*Glyptograptus*' cf. *teretiusculus* and *Normalograptus* cf. *brevis*. These assemblages are thought to represent part of the *Nemagraptus gracilis* Biozone or the base of the *Diplograptus multidens* Biozone (Table 2). Poorly preserved graptolites from low in the formation, east of Llanegryn [6069 0589], include doubtful examples of *Amplexograptus arctus*, *Cryptograptus*, *Lasiograptus*, *Orthograptus* cf. *calcaratus* s.l. and *Pseudoclimacograptus scharenbergi*. The possible presence of *A. arctus* suggests the *multidens* Biozone. Jones (1933, p. 163) recorded *Dicellograptus* from about the same horizon. Other collections from workings along Ffordd Ddu [6251 0885 and 6361 1020], which include *Lingulella*, *'Glyptograptus' euglyphus* and *Pseudoclimacograptus*?, are biostratigraphically less definite.

Acritarchs from the Ty'r Gawen Mudstone Formation, south-east of Bwlch-coch [7482 1540], include *Micrhy-*

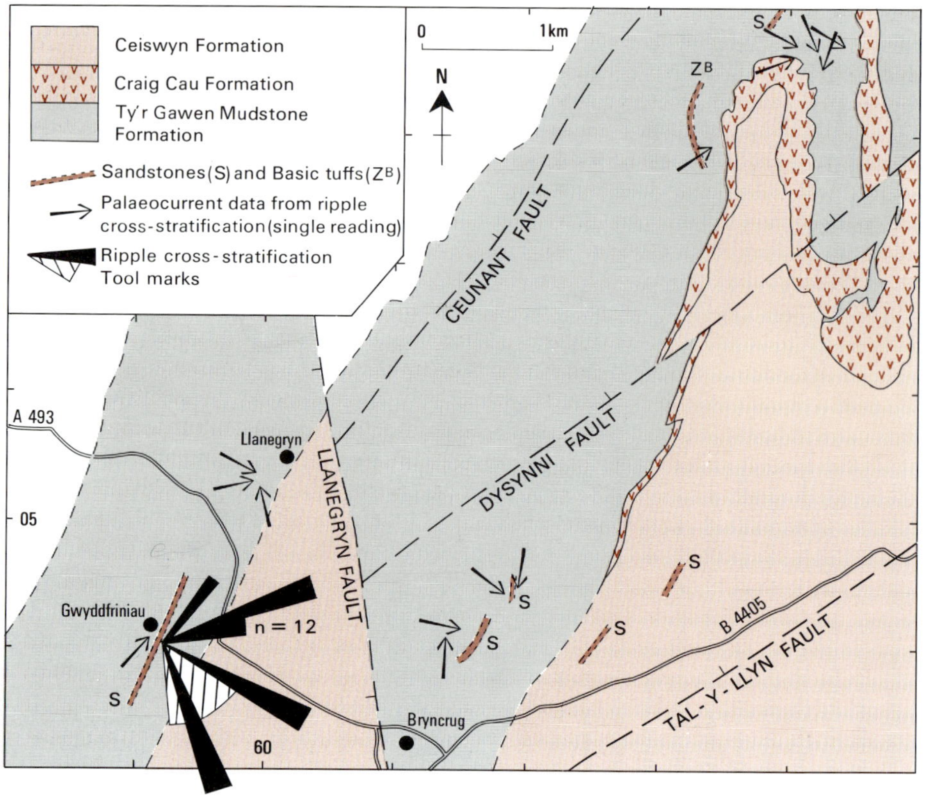

Figure 18 Palaeocurrent data from the Ty'r Gawen Mudstone Formation around Llanegryn.

stridium aremoricanum which does not occur above the Llandeilo (Turner, 1985), implying this age for at least part of the formation.

CONDITIONS OF DEPOSITION

The ironstone and associated mudstones of the Fron Newydd Member suggest a period of low sedimentation rate during Llandeilo or early Caradoc times. Ancient oolitic ironstones have few modern analogues and their origin is contentious (e.g. Curtis and Spears, 1968; Kimberley, 1974, 1979; Taylor, 1990). It is unlikely that they were deposited in an environment similar to that in which modern carbonate ooliths develop (e.g. Newell et al., 1960), as there is no evidence of shallow water conditions. Furthermore, the preservation of delicate concentric banding within the ooliths of the Fron Newydd Member implies that the chamosite was primary (cf. Hallimond, 1925; Taylor, 1949; Odin and Matter, 1981), rather than a diagenetic alteration of calcite. The associated basic volcanic rocks could well have been the source of the iron (cf. Strahan et al., 1920; Trythall et al., 1987) and it is probable that the ooliths were transported into the district from shallower water. Trythall et al. (1987) considered that the thick phosphatic mudstones were probably deposited farther from the source of the ooliths. The phosphate nodules may have developed in situ, but the restriction of most ooliths within them suggests that a phosphatised hardground was broken up and resedimented, probably by debris flows. This interpretation is consistent with the other evidence of slumping in the member.

West of the Llanegryn Fault, ironstone deposition was preceded by late Llanvirn uplift and erosion. However, because of the large Mesozoic/Tertiary displacement on the fault, the contrast across it cannot be interpreted easily and a pre-Mesozoic expression of the fault (cf. Kokelaar, 1988) remains unproven. It is possible that rather than being related to the fault, the changes reflect uplift of, and onlap onto, an uplifted Harlech Dome.

Localised slumping in the mudstone sequence above the Fron Newydd Member implies instability and deposition from debris flows. Turbiditic sedimentation is apparent only in the west of the district, where sandstones near the top of the formation are interpreted to be the lobes of northerly or north-westerly derived submarine fans. The petrographical assemblages of the sandstones are diverse and imply derivation from a mixed terrain of volcanic rocks and metamorphic basement, possibly the Irish Sea Landmass (e.g. Brück et al., 1979). Thin tuffs and tuffites underlying the Craig Cau Formation near Llyn Cau possibly represent reworking of pyroclasts in the precursor phase of the major activity involved in emplacement of the Craig Cau Formation.

Pen y gadair Volcanic Formation

The Pen y gadair Volcanic Formation, previously the Upper Basic and Pen y Gader groups (Cox, 1925; Cox and Wells, 1927), mainly comprises pillowed, massive and autobrecciated basalt lavas with intercalated crystal tuffs and few acid, ash-flow tuffs. Rare irregular bodies of basaltic volcanic breccia are present but difficult to interpret. The formation interdigitates with the Ty'r Gawen Mudstone Formation and is thickest, about 240 m, between Cadair Idris and Cross Foxes (Figure 7). Locally, as south of Cross Foxes, the Ty'r Gawen Mudstone Formation is absent and the formation rests directly on the Llyn y Gafr Volcanic Formation. The most westerly development of basalt occurs on Tyrrau Mawr [6745 1348], where the sequence is mainly of acid tuff, but the crystal tuffs persist to Ffordd Ddu [6483 1277], a few kilometres further west. A marked lateral facies change, similar to that recognised in the Llyn y Gafr Volcanic Formation, occurs in the vicinity of Cross Foxes, where basalts wedge out eastwards into volcaniclastic rocks.

On Tyrrau Mawr, the formation, 95 m thick, comprises three massive, acid tuffs with intercalated mudstones and tuffaceous sandstones, overlain by a basaltic breccia (Figure 15). The acid tuffs, up to 26 m thick, include clasts, up to 0.25 m long, of bedded tuff and mudstone, and feldspar crystals, up to 50 per cent. In thin section, devitrified and quartzose recrystallised shards can be distinguished in a sericite/chlorite matrix (E61596-8, E61600). The two lower acid tuffs contain a bedding-parallel welding foliation, for much of their thickness, and grade up into thinly bedded, finer-grained tuff. The basaltic, volcanic breccia forms a massive, lenticular deposit [6725 1311] at least 30 m thick, with a planar base but an irregular upper surface. It wedges out into mudstones within 300 m to the south-west. To the north-east, it is poorly exposed and appears to pass laterally into a massive basalt/fine-grained dolerite intrusion/extrusion. The breccia comprises angular blocks, up to 0.2 m long, of vesicular, carbonated and chloritised basalt in a matrix of fine-grained, basaltic fragments. The coarsest matrix components are scoriaceous and these grade into finer shards (E61584). Rare acid tuff fragments also occur (E65204).

In the vicinity of Penygadair, the summit of the Cadair Idris escarpment, the formation comprises a basal crystal tuff, about 10 m thick, separated by an intercalation of Ty'r Gawen Mudstone Formation (95 m thick) and a large microgranite intrusion, from an upper sequence of 200 m of pillowed and massive basalts (Figure 15). East of Llyn y Gadair [7113 1371], the basal tuff comprises massive, clast-rich tuff, 2 m thick, overlain by finer-grained, planar bedded-tuff with some low-angle cross-stratification. Highly altered, feldspar crystals, 1–3 mm in length, are abundant (30–50 per cent) and the basal part includes clasts, up to 60 mm in diameter, of basalt, basic tuff and acid tuff (E66798). The mudstones immediately above and below the crystal tuffs contain many tuffaceous layers. For example, east of Llyn y Gadair [7113 1371], numerous fine-grained, tuffaceous turbiditic sandstones are interbedded with mudstones in the 15 m below, and 10 m above, a crystal tuff. Some of the sandstone beds are up to 0.1 m thick, feldspar crystal-rich, normally graded, faintly planar laminated, and include chloritised basic shards (E66799).

The formation is well exposed around, and west of, Penygadair [6920 1235 to 7110 1310], the type locality, and a detailed sequence has been distinguished (Figure 19). The section is dominated by rusty weathered, basalt lava flows, mainly 2–15 m thick. Flow contacts are well defined where thin tuffs and mudstones intervene, but, elsewhere, they are less easily distinguished by rusty weathered, oxidised layers and zones of chilled, pale green basalt. Most lavas are either massive or pillowed, but the thicker flows commonly display a massive base, with incipient columnar jointing, and a pillowed top. The pillows in individual flows commonly decrease in size upwards, from 2 m across to less than 0.2 m and in places display concentric and radial fractures [7052 1264, 7007 1257, 7105 1280] (Plate 8a). Most pillows contain quartz-filled vesicles and are closely packed with little or no interpillow sediment. Rare lava tubes, up to 10 m across, infilled with flow banded basalt, have also been distinguished [7160 1302]. Locally [7035 1280], hyaloclastites, up to 5 m thick, infill depressions on the tops of flows. At Penygadair, the basalts include two intercalated acid, ash-flow tuffs, each up to 15 m thick, the lower of which can be correlated with the lowest tuff at Tyrrau Mawr, 10 km to the west (Figure 15). They comprise massive, welded tuff overlain by bedded tuffs with rare large-scale, low-angle cross-stratification and channels [7104 1290, 7110 1307, 7041 1263, 7026 1266]. The tuffs are composed of a devitrified and recrystallised quartzose aggregate with sodic plagioclase crystals, shards and pumice, and, locally, a eutaxitic foliation can be distinguished (E62990, E62992, E66207, E66337).

To the east of Penygadair, the upper acid tuff wedges out but the basalts and crystal tuffs thicken and, in the vicinity of Mynydd Moel, the intercalated mudstones of the Ty'r Gawen Mudstone Formation thin markedly (Figure 15). Irregular bodies of basalt and fine-grained dolerite, probably high-level intrusions, are also common. The lower acid tuff locally contains subrounded basalt clasts up to 3 m in diameter [7374 1385], and thins northeastwards to less than 10 m. The basalt lavas are mainly pillowed [7331 1357] with massive lower parts, but some are pillowed or massive throughout. For example, near Gau Graig [7409 1404], the uppermost lava, about 10 m thick, which is massive, with a highly vesicular top, 4 m thick, can be traced for about 500 m along strike. Hyaloclastites, up to 2.5 m thick, are interbedded with the basalts [7409 1403, 7511 1484] and comprise angular, chloritised and ferritised, 'glassy' basalt clasts, up to 5 mm in diameter, with isolated pillows and nonvesicular and feldsparphyric pillow fragments (E65982, E65760). Basic tuffs, up to 2.25 m thick, infill shallow depressions on the tops of some lava flows [7444 1428, 7439 1425].

Between Gau Graig and Cross Foxes, the formation (200 m thick) is less well exposed and correlation is locally complicated by a basaltic breccia. At the base of the formation [7518 1575], crystal tuffs and tuffites, up to 35 m thick, are interbedded with basalt lavas [7583 1604]. The crystal tuffs closely resemble those at the base of the laterally equivalent Benglog Formation (Plate 7b), and are poorly sorted with up to 50 per cent feldspar crystals, subangular clasts, up to 0.5 m across, of vesicular and feldsparphyric basalt, mudstone, crystal tuff, basic tuff, acid volcanic rock and quartzose sandstone. The euhedral to fragmented feldspar crystals, which are up to 5 mm across, are commonly highly altered (E65160, E65190).

North-west of Gwerngraig [7512 1553 to 7545 1597], the sequence includes a problematic basaltic breccia. Be-

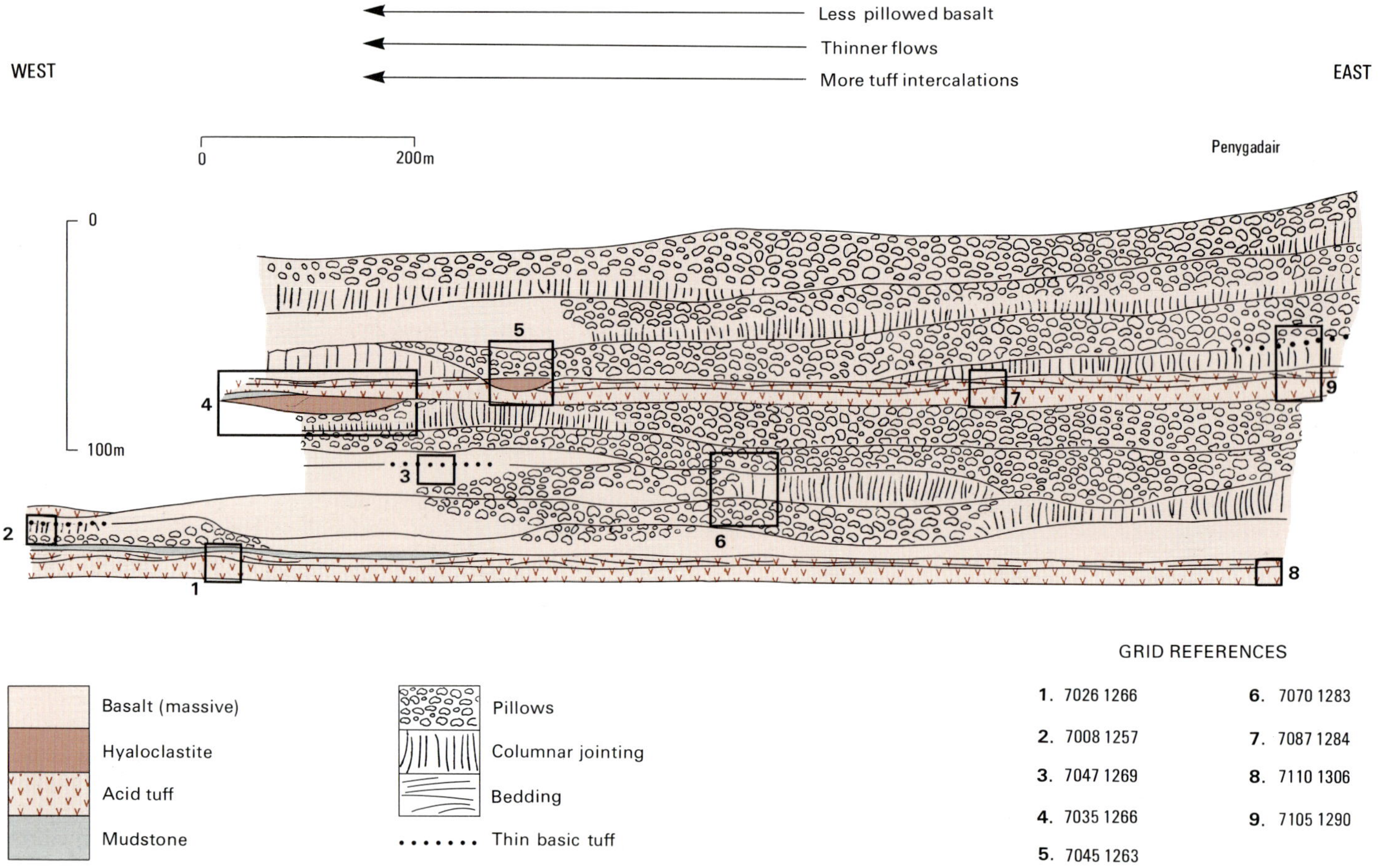

Figure 19 Details within the Pen y gadair Volcanic Formation west of the summit of Cadair Idris (Penygadair) [7008 1257 to 7105 1290]. The numbered boxes, with grid references, indicate specific localities.

cause of poor exposure, it is difficult to distinguish its overall form, but it has a dome-like geometry, with a maximum thickness of about 200 m, and an approximately concordant base. Its relationships with the adjacent strata are uncertain; to the west it is intruded by the 'Pen-y-Gader dolerite' and to the east it appears to interdigitate with a sequence of pillowed basalt, basic tuff, and intrusive dolerite. The breccia is massive, rarely bedded [7525 1549], poorly sorted, and consists mainly of abundant angular fragments, 5–250 mm across, of chilled, nonvesiculated basalt, comprising sodic plagioclase phenocrysts in a chloritised glass matrix (E65147), in a finely comminuted chloritic matrix with large, cuspate, basaltic glass fragments (E65151). Locally [7543 1573], it also includes fragments of subophitic dolerite (E65151-3), basic tuff and mudstone.

Petrography

The basalt lavas of the Pen y gadair Volcanic Formation are petrographically indistinguishable from those of the Llyn y Gafr Volcanic Formation. Massive lavas are generally coarser grained, commonly with subophitic textures (E65209, E65754), whereas the pillowed basalts are finer with feldspar laths, locally forming radiating aggregates, in a glassy mesostasis (E66340-1, E65150, E65181, E65981) (Plate 7a). Some coarsely textured massive basalts [7188 1266, 7181 1264] are composed of abundant euhedral to suhedral feldspar crystals, up to 1 mm in length, with an intersertal, finer-grained basalt mesostasis (E62621, E62982).

Biostratigraphy

At Llyn y Gadair [7109 1374], mudstones, about 40 m above the Fron Newydd Member, have yielded abundant '*Glyptograptus*' *teretiusculus*, with rarer *Climacograptus* sp., *Dicranograptus* [large species], *Diplograptus foliaceus, Lasiograptus costatus*? and *Pseudoclimacograptus* sp. (Figure 14 d–f). When compared with the ranges of graptolites in the Shelve Inlier (Hughes, 1989) and southern Sweden (Nilsson, 1977), the overlap of the ranges of *'G.' teretiusculus* and *L. costatus* indicates the top of the *gracilis* Biozone or the base of the *multidens* Biozone (Table 2). Specimens in Cox's collection from the west side of Tyrrau Mawr [about 6732 1332] may be from the same formation; they are referred to *Amplexograptus*? *molestus* (Figure 14g, h), *Normalograptus*? and *'Glyptograptus'* sp., suggesting the *gracilis* or *multidens* Biozone. A specimen of *Pseudoclimacograptus isknos*?

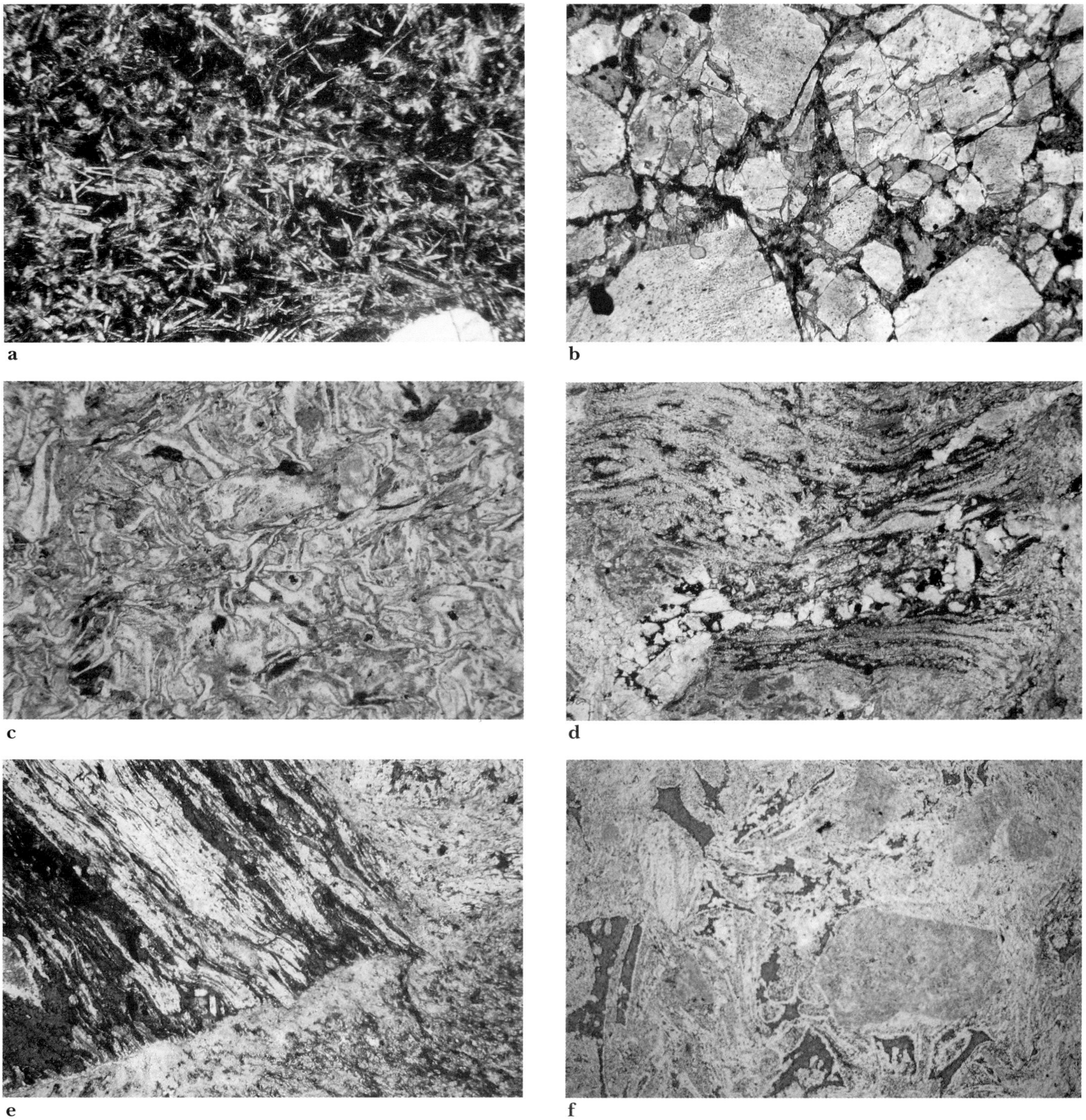

Plate 7 Photomicrographs of Aran Volcanic Group lithologies from formations above the Fron Newydd Member.

a) Basalt. Pen y gadair Volcanic Formation (E66341). Southwest of Penygadair [7065 1283]. Field of view about 7 mm wide.
b) Crystal tuff with numerous albite-oligoclase crystals. Benglog Formation (E65022). East of Tyddyn Du [7685 1508]. Field of view about 6 mm wide.
c) Non-welded vitroclastic tuff/tuffite from the lower part of the Craig Cau Formation (E65750). Gau Graig [7567 1491]. Field of view about 4 mm wide.
d) Welded acid tuff from the lower part of the Craig Cau Formation (E62630). East of Llyn Cau [7242 1192]. Field of view about 4 mm wide.
e) Welded, lithic tuff with large pumice fragments, from the upper part of the Craig Cau Formation (E66206). Minffordd [7363 1184]. Field of view about 6 mm wide.
f) Weakly welded acid tuff (E66220). Craig Cau Formation. Craig Cau [7093 1240]. Field of view about 6 mm wide.

Plate 8a Pillowed basalts. Pen y gadair Volcanic Formation. Southwest of Penygadair [7095 1298].

Plate 8b Block-rich tuff at the base of the Craig Cau Formation. Bwlch Cau [7036 1227].

which was collected from above the Pen y gadair Volcanic Formation on Tyrrau Mawr [6749 1307], though doubtful, is suggestive of the *multidens* Biozone (Zalasiewicz, 1992).

Conditions of deposition

The Pen y gadair Volcanic Formation reflects an episode of effusive and explosive, subaqueous basaltic volcanism similar to that of the Llyn y Gafr Volcanic Formation. The wedging out of basalts in the vicinity of Cross Foxes, in a similar position to the basalt–tuff transition in the Llyn y Gafr Volcanic Formation, suggests little change in the position of the volcanic centre(s) and basin geometry. The formation was emplaced after a period of volcanic quiescence marked by deposition of the Fron Newydd Member ironstone and the lowest mudstones of the Ty'r Gawen Mudstone Formation. The acid ash-flow tuffs, some welded, indicate contemporaneous explosive acid volcanism. The source is not recognised, but the westward thinning of the basalts and tuffs suggests transport broadly in that direction. It is possible that the Gwerngraig basaltic breccia is the result of an explosive reaction from high-level intrusion into wet sediment, but cuspate basalt inclusions, the scarcity of interstitial sediment, and the incorporation of dolerite blocks suggests accumulation close to a vent.

Plate 8c Disturbed mudstones forming prominent ridges dipping at approximately 40° to the south (left). Ceiswyn Formation. Nant Ceiswyn [7885 1490].

Benglog Formation

The Benglog Formation (Allen and Jackson, 1985), broadly equivalent to the Pen y gadair Volcanic Formation (Figure 7), crops out in the north-east of the district and comprises crystal tuffs, similar to those at the base of the Pen y gadair Volcanic Formation, and interbedded dark grey mudstones. West of the Bala Fault the formation, about 90 m thick, crops out north of Cross Foxes but is poorly exposed. East of the Bala Fault it is up to 250 m thick, and crops out between Tyddyn Du and Bwlch Oerddrws.

Around Bwlch Oerddrws [7940 1705], the formation, about 200 m thick, comprises crystal-rich and blocky tuffs and tuffites with intercalated mudstones, locally [7895 1630] with basalt clasts up to 80 mm in diameter, and thin tuffaceous sandstones. The coarse pyroclastic and tuffaceous rocks are massive, locally reverse graded at the base, and commonly normally graded at the top (Figure 16). The crystal tuffs contain abundant (20–50 per cent) euhedral, sericitised, sodic plagioclase crystals, up to 5 mm long (E66773), few quartz crystals (mostly 5–10 per cent) (E66773–4), and angular to subangular clasts, 5 mm to 0.5 m across, of basalt (E66771), mudstone, bedded tuff and acid volcanic rock (Plate 7b). Locally [7786 1585], the crystal tuffs contain mudstone fragments which are indented by adjacent feldspar crystals, indicating incorporation in a partly lithified state (E65037). The matrix of the crystal tuffs and tuffites is commonly chlor-itic and devitrified shards are rarely distinguished (E66771–2). Tuffaceous sandstones are abundant in the lowest 40 m of the formation (Figure 16) and comprise angular to subangular grains, 0.5–2 mm in diameter, of feldspar (over 50 per cent), lithic fragments, mainly basalt and acid tuff, quartz and mica (E66764–5).

Biostratigraphy

Mudstones between Nant Ffridd-fawr and Bwlch Oerddrws [7888 1634] yielded the graptolites *Climacograptus* sp., *Diplograptus foliaceus* (Figure 14a–c), *'Glyptograptus' teretiusculus* and *Normalograptus* cf. *brevis*, which indicate the *gracilis* Biozone or lower *multidens* Biozone (Late Llandeilo or Early Caradoc; Table 2).

Conditions of deposition

The crystal tuffs are interpreted as pyroclastic flows and debris flows resulting from explosive, subaqueous, eruptions. The variable proportion of mudstone matrix, and variety of lithic fragments, probably reflects incorporation of sediment during transport or eruption through variably lithified sediment. The euhedral shape of most feldspar crystals suggests that they are pyroclasts rather than the product of erosion of crystal-rich basalts and their high concentration implies derivation from a crystal-rich magma. The only comparable feldsparphyric, basic igneous rock within the district occurs at the base of the Llyn y Gafr Volcanic Formation. However, feldspar phenocrysts and crystals are widespread in the acid tuffs, rhyolites and microgranites of the Aran Volcanic Group and Zalasiewicz (1981) suggested that intrusive/extrusive quartz latites in the Arenig area were an important source of feldspars.

Craig Cau Formation

This formation is equivalent to the Upper Acid and Craig y Llam groups of Cox and Wells (1921), the Craig y Llam Formation of Ridgway (1975) and the Perfeddnant Ashes of Jehu (1926). It is also the lateral equivalent of the Llyn Conwy Formation of Arenig (Zalasiewicz, 1992). Cox (1925) and Davies (1959) described many of the rocks of the formation as rhyolites, keratophyres and trachytes, but the main component is acid tuff. Rhyolites, which are petrographically similar to the acid tuffs, occur in many places, particularly around Craig Cwmrhyddfor; generally, their form is uncertain, but some are demonstrably intrusive.

The formation is about 400 m thick over much of the central and eastern parts of the outcrop, but it thins west of Castell-y-bere and wedges out near Bryncrug (Figure 7). It is well exposed around Craig Cau (the type section [7100 1217]), at Bird Rock, in the Pennant and Craig Ysgiog anticlines, and on the high ground around Craig y Llam and Bwlch Oerddrws. Where the formation overlies mudstone, its base is well defined, but in the north-east of the district, around Bwlch Oerddrws, it overlies crystal tuffs of the Benglog Formation and the base is less clearly distinguished.

Subdivision of the formation is made difficult by widespread syndepositional disruption and the scarcity of persistent sedimentary intercalations. The observation of Fearnsides (1905), on equivalent strata at Arenig, '.....what may appear to be good mappable lines at one place usually lapse into a maze of similarities within a very few hundred yards along the outcrop' is also pertinent to the Craig Cau Formation. However, between the north-east edge of the district and Mynydd Pennant, a broad subdivision of the formation is possible into a lower sequence of tuffs, tuffites and unfossiliferous mudstones, and an upper, single, acid welded tuff (Figure 20). In much of the outcrop, the lower sequence comprises at least three ash-flow tuffs and tuffites, locally with thin interbedded mudstones. However, apart from a block-rich, acid tuff at the base, correlations for any significant distance along strike are difficult. There is also widespread evidence of slumping and loading within the lower sequence: tuff rafts and slump folded tuff beds are common and the bases of more continuous tuff beds are loaded into, and penetrated by flames of, the underlying mudstone.

At Craig Cau [7040 1227], the basal, block-rich tuff, 20–40 m thick, comprises clasts, 30 mm to 1 m long, mainly of vesicular basalt, with fewer of dolerite, basic tuff, pumice and, most rarely, rhyolite, in a dark green, muddy matrix with scattered feldspar crystals up to 4 mm long (Plate 8b; Figure 20). Some of the basalt clasts resemble broken pillow rinds. Although there are local concentrations of large clasts in the basal 15 m, the sorting is generally poor and the clast and mudstone content is very variable. The upper part of the tuff is largely clast free and is distinctly more acid in appearance. In thin section, the matrix is a strongly recrystallised quartzo-feldspathic mosaic, rich in feldspar microlites and clusters of euhedral sodic feldspar crystals, with rare needles of apatite (E65797, E66214).

At Craig Cau the basal tuff is overlain by a disrupted sequence of tuffites, pumice-rich acid tuffs, mudstone debris-flow deposits, and a few laterally impersistent, massive mudstones, 1–6 m thick. Impersistent pods of possibly extrusive fine-grained basalt (E62637, E62635) also occur a short distance along strike to the east, at the edge of Cwm Cau [7246 1218]. The debris-flow deposits, up to 25 m thick, comprise scattered, poorly sorted angular fragments of acid volcanic rock, up to 0.5 m in length, within a massive mudstone matrix. The tuffs and tuffites, 10–50 m thick, are massive, weakly cleaved and with many albite-oligoclase feldspar prismatic crystals up to 5 mm long. Many contain silicified mudstone 'rip-up' clasts, up to 1 m in length. Weathered surfaces of many of the tuffs indicate a welding foliation. In thin section, most textures are obscured by quartzofeldspathic recrystallisation, but unwelded vitroclastic textures and welding foliation are preserved locally (Plate 7d and f).

At Bwlch Oerddrws [7950 1700], a massive, acid ash-flow tuff, 75–170 m thick, at the base of the formation, is probably the lateral equivalent of the Craig y Ffynnon Formation to the north-east (Dunkley, 1978, 1979). It is overlain by a sequence of slumped mudstones with few intercalated tuffs and basalts, which is broadly equivalent to the Pistyllion Formation (Dunkley, 1978, 1979). The mudstones are massive and contain isolated, subangular to subrounded fragments of tuff and basalt [7815 1537, 7973 1637]. The anomalous total thickness of the formation in this area, about 600 m, is probably the result of the repetition of the lower part by stacking of slump sheets or by overturned slump folds.

The correlation of the sequence from Bwlch Oerddrws to Cadair Idris, across the Bala and Tal-y-llyn faults, is uncertain, mainly because of poor exposure. However, the basal acid tuff at Bwlch Oerddrws may be a less contaminated, lateral equivalent of a muddy, acid tuffite, 20–30 m thick, which occurs at the base of the formation on Craig Cwmrhyddfor and at Gau Graig [7442 1415]. The tuffite comprises variable proportions of highly cuspate shards, more than 0.5 mm across, within a mudstone matrix (Plate 7c).

The contact between the lower part of the formation and the upper, welded tuff is clearly distinguished where mudstones or bedded tuffs intervene, as at Craig Cau [7018 1206], where the tuff overlies 13.5 m of bedded tuffs (Figure 20). These include massive acid tuffs, about 1.5 m thick, which grade up into planar-laminated tuff. The lowest beds contain normally graded, rounded basalt fragments up to 0.15 m in diameter. However, to the east, the bedded tuffs are progressively overstepped by the welded tuff and within a few hundred metres [7042 1209] are reduced to 5.5 m and then cut out, so that the welded tuff overlies pumice-rich, muddy tuff at Craig Cwm Amarch [7105 1214]. Farther east, the base of the welded tuff is less clear. In a distant view of the cliffs on the south side of Llyn Cau [7150 1210], it forms a bench-like feature, probably caused by a few metres of mudstone, but on the accessible outcrop its position is difficult to determine.

The upper ash-flow tuff is probably the lateral equivalent of the Aran Fawddwy Formation (Dunkley, 1978, 1979). It is present throughout the outcrop and is the sole representative of the formation west of Mynydd Pennant. In its most westerly exposure, on Foel Wyllt [6264 0460], it comprises 2 m of weakly cleaved, muddy tuff, with shards between 0.25 and 0.5 mm in length (E67085), overlain by 2 m of laminated tuff. A few kilometres to the north [6331 0700], the ash-flow tuff, up to 50 m thick, is welded and it thickens to 180 m on Cadair Idris (Figure 20). At Bwlch Oerddrws the ash-flow tuff is about 200 m

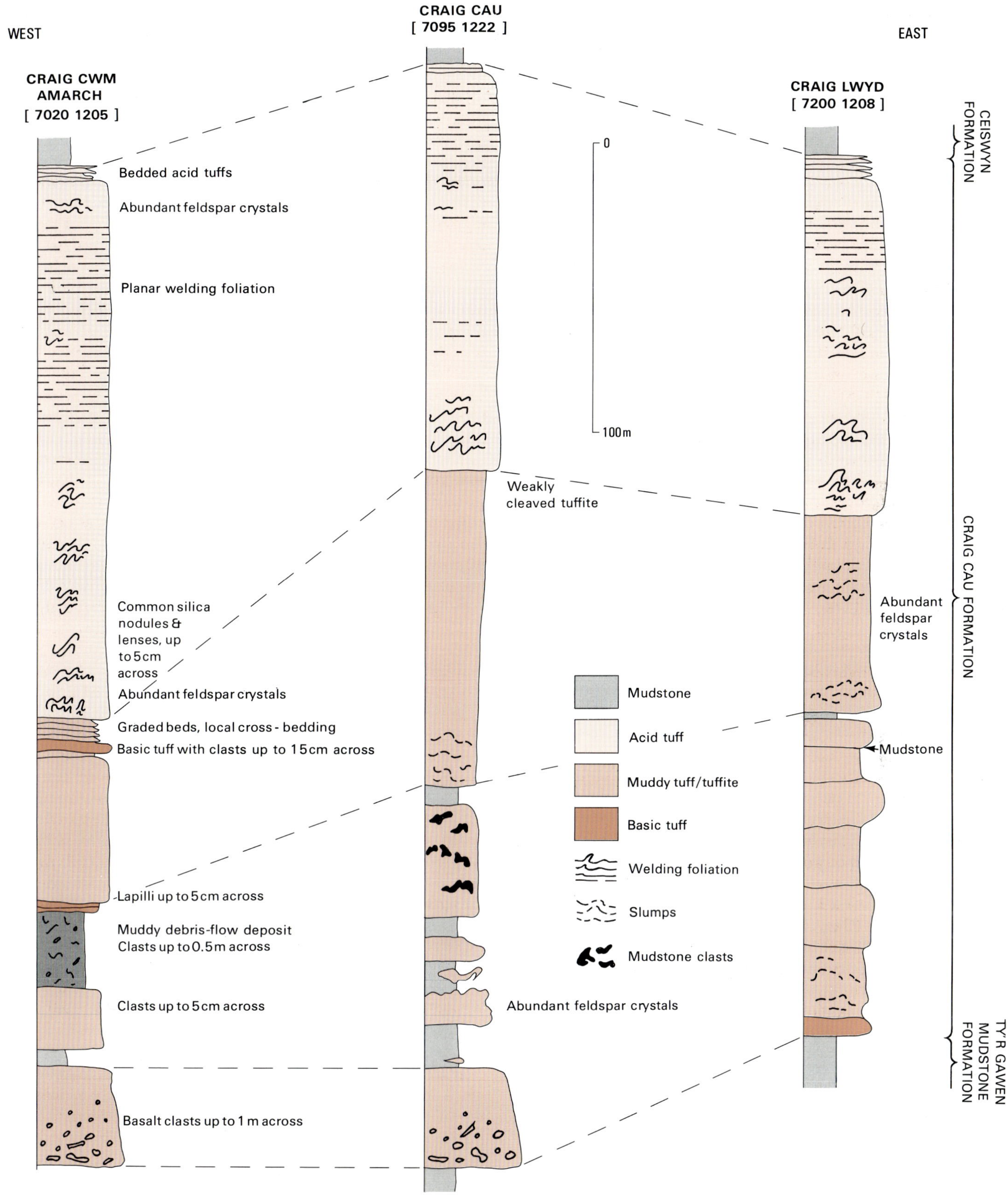

Figure 20 Simplified graphic logs of the Craig Cau Formation, approximately 1 km apart.

thick; the lowest 120 m of nonwelded tuff, rich in lithic fragments, is overlain by 80 m of strongly welded tuff.

In most exposures, the ash-flow tuff weathers white or pink and contains scattered euhedral, sodic plagioclase laths up to 4 mm in length. Quartz veins and small (up to 10 mm) quartz nodules, flattened in the cleavage, are widespread. Cooling joints are common and the welding foliation is accentuated by variable recrystallisation in adjacent bands. Pumice and lithic fragments, mostly acid tuff and rhyolite, are also common locally, and, in places, dark green pumice fiammé form almost 50 per cent of the tuff, for example near Minffordd [7263 1184] (E66206; Plate 7e). Rafts of laminated acid tuff, up to 2 m long, occur in the top 10 m of the ash-flow tuff at Castell-y-bere [6678 0856] and Bird Rock [6497 0718].

The intensity and orientation of the welding foliation in the upper ash-flow tuff show systematic variations. Near Craig Cau (the type section) [7100 1217], the lowest 2–3 m comprise nonwelded tuff crowded with albite/oligoclase crystals and well-developed quartz nodules (E66219). Between Craig Lwyd [7205 1211] and Craig Ysgiog [6806 1037], the welding foliation in the basal zone is rheomorphosed into mesoscale folds and closely resembles flow-folded rhyolite. Above, the welding foliation becomes progressively less contorted and is orientated close to bedding near the top of the tuff. A distinct, slightly undulating foliation is common in the upper part of the tuff, for example at Mynydd Pennant [6596 0905] and Craig Ysgiog [6816 1007], and at Bird Rock [6468 0608] the foliation is randomly oriented in the main body of tuff, but planar and bedding-parallel in the uppermost 10–15 m. In thin section, eutaxitic foliation is only rarely discernible (E62624, E66355), as fine textural details are commonly obscured by recrystallisation and silicification. The effects of these processes are particularly intense in the rheomorphosed tuffs and these rocks are difficult to distinguish from rhyolites (E61956–7, E66221, E62625).

Syndepositional disruption of the tuff is apparent in the west part of the outcrop, where the base of the ash-flow tuff is locally irregular, lobing into the mudstone substrate. In places, pods of tuff 4–100 m in diameter, very similar to those of the Capel Curig Volcanic Formation in north Snowdonia (Francis and Howells, 1973; Howells et al., 1991), are detached and surrounded by mudstone. Near Dysefin, in the vertical eastern limb of the Bird Rock Anticline, tuff pods detached from the base of the tuff are elliptical in plan view. The larger tuff pods have smooth, rounded surfaces with a 1–2 m wide margin of yellow/pale green, sericitised tuff with numerous concentrically zoned quartz/chlorite nodules, up to 0.1 m in diameter. Smaller tuff bodies are generally composed entirely of sericitised tuff and typically show complex interdigitating contacts with the mudstone matrix.

Pods detached from the main tuff are widespread, but at Bird Rock the tuff appears to be completely breached along a north-east–south-west zone, 100–200 m wide. The tuff pods exposed within this steep sided zone, which may represent the trace of a syndepositional fault, are cigar-shaped in plan and aligned parallel to the margins of the breached zone.

The ash-flow tuff is overlain by a bedded sequence, 0.5–10 m thick, of laminated, very fine-grained acid tuff and medium to thickly bedded, acid tuffites with very thin intercalated mudstones which form the prominent dip slopes along much of the outcrop, for example Castell-y-bere, Mynydd Pennant and Craig Lwyd. The laminated tuffs are mainly alternations of silt and mud grade components in beds between 1 mm and 30 mm thick and in thin section they are seen to be a mosaic of quartz and feldspar after devitrified and recrystallised ash (E67086). The tuffites form structureless, locally graded beds up to 1 m thick and contain shards and feldspar fragments within a mudstone matrix (E61954). West of Cadair Idris, as at Llanllwyda [6501 0750], the bedded sequence, 0.5–5 m thick, is entirely of laminated tuff. In contrast, east of Cadair Idris laminated tuffs are rare in the bedded sequence and their cumulative thickness seldom exceeds 1.5 m.

Conditions of deposition

The formation represents the final and most explosive phase of volcanism of the Aran Volcanic Group. The lower sequence reflects widespread remobilisation of pyroclastic debris by sedimentary processes and the slumped and disrupted strata indicate that the instability recognised in the underlying Ty'r Gawen Mudstone Formation continued. Although the basal, blocky tuff was probably the product of an acid pyroclastic flow, the high proportion of basalt blocks within it indicates contemporaneous basalt eruption and incorporation into the flow. Broadly contemporaneous basic volcanism is also suggested by the occurrence of small pods of basalt, either as extrusions or high-level intrusions, which occur at the same stratigraphical level as the blocky tuff in Llyn Cau, about 1.5 km along strike.

Widespread welding, nodule development and the lack of depositional breaks indicates that the acid tuffs of the formation were deposited from a series of large, hot pyroclastic flows. The rheomorphosed foliation near the base of the highest ash-flow tuff indicates postdepositional and pre-lithification movement. Interaction of hot tuff with water saturated substrate facilitated circulation of water and initiated the processes of devitrification. Collapse of the tuffs into the mudstone substrate was coincidental. Further collapse was probably caused by a combination of fluidisation and unequal loading in a subaqueous setting (cf. Kokelaar, 1982; Howells et al., 1985, 1991). The relative lack of reworking of the upper surface of the tuff indicates emplacement in deep water. Most of the laminated tuffs probably represent fine ash elutriated from the top of the pyroclastic flows during transport which settled out from suspension in the water column.

It has been suggested that the Cadair Idris microgranite, a late-stage intrusion, lies close to the eruptive centre of the Craig Cau Formation (Cox and Wells, 1927; Davies, 1959). The boss-like, rhyolitic southern margin of the intrusion, which may represent the transition from intrusion to extrusion, transects the formation at Craig Cwmrhyddfor and is overlain by the Ceiswyn Formation. Similar intrusive/extrusive relationships occur at another microgranite at Mynydd Pennant and they may be representative of the same magma source. However, there is no evidence of a caldera structure and the wide distribution of the formation, and its lateral variation, about the east and south sides of the Harlech Dome suggests that it evolved from an eruptive centre(s) outside the district.

POST-ARAN VOLCANIC GROUP STRATA

The Aran Volcanic Group is overlain conformably by a thick succession of Caradoc and Ashgill mudstones with few siltstones, sandstones and conglomerates. The strata occupy about one third of the district and crop out between Cribin Fawr, in the east, and the coast at Tywyn. In addition, they form small, faulted inliers either side of Machynlleth. In upward succession, the sequence comprises: the Ceiswyn, Nod Glas, Broad Vein, Narrow Vein and Garnedd-wen formations (Table 1). These subdivisions closely follow those of Pugh (1923) and Jehu (1926) and are derived largely from quarrying names, such as 'Nod Glas' (Blue rock), 'Y Faen Lydan' (Broad Vein) and 'Y Faen gul' (Narrow Vein) (Davies, 1878).

During Caradoc–Ashgill times, the district probably lay in one of the deeper parts of the Welsh basin. To the east, the sequence thins and includes carbonate-rich mudstones, discrete limestones, unconformities and abundant shelly faunas (Pugh, 1929; Bassett et al., 1966; Lockley, 1980a, 1980b). Sparse palaeocurrent data indicate sedimentary input from the north, north-west, north-east and south-east at different times, and possibly indicate that the district lay close to the axis of the basin. Much of the mudstone-dominated turbidite sequence represents a period of steady subsidence. However, during the late Ashgill (Hirnantian) glacioeustatic regression, there was widespread slumping and incursions of shelf-derived sandstones deep into the basin (e.g. Cave, 1979; Woodcock and Smallwood, 1987).

Ceiswyn Formation

The Ceiswyn Formation, mainly mudstone and siltstone, is the lateral equivalent of the Nant Hir Mudstones or Lower Bala Group of the Bala district (Bassett et al., 1966). To the east of the district, between Bala and Llanymawddwy, greater lithological variation allowed Pugh (1929), Bassett et al. (1966) and Lockley (1980a) to subdivide the equivalent strata. However, within the district, only one new division, the Craig Hen-gae Member, is recognised (Pratt, 1990). This is the Upper Ceiswyn Beds of Pugh (1928), equivalent to the Gelli-grin Formation of the Bala district (Bassett et al., 1966; Lockley, 1980a).

West of Tal-y-llyn Pass, thickness measurements are difficult because of problems in correlation across the Tal-y-llyn Fault and tight folding of the sequence on the north side of the fault. The maximum exposed thickness, in a river section south-east of Abergynolwyn [6800 0680], is about 1300 m. However, east of Tal-y-llyn Pass the succession is complete and the thickness declines from 1550 m at Craig Hen-gae [7580 1023] to 1400 m at Nant Ceiswyn [7980 1424], about 6 km to the north-east. East of the district, at Dinas Mawddwy, Pugh (1929) calculated 1330 m of equivalent strata. The type section, at Nant Ceiswyn, is a combined stream and forestry track section [7878 1537 to 7980 1424] and lies within the lower 900 m (Figure 21).

There is little indication of reworking of the underlying Craig Cau Formation and over much of the district the basal 50 m are well-cleaved, turbiditic mudstones, with dark grey, laminated (hemipelagic) mudstone beds and widely spaced, thin siltstones, 10 mm thick. These beds are well exposed in slate quarries at Perfeddnant [6301 0535], Bird Rock [6340 0707], Fron Fraith [7630 1321] and Cribin Fawr [7975 1592]. Elsewhere, for example at Mynydd Pencoed [7062 1184], the Craig Cau Formation is overlain by 4–5 m of disturbed, massive mudstones. At Nant Ceiswyn, Fron Fraith, Bird Rock and Castell-y-bere much of the lowest 350 m comprises disturbed mudstone (Figure 21).

The bulk of the Ceiswyn Formation comprises thinly bedded, turbiditic silty mudstones, 0.01–0.2 m thick, locally with thin, 10–20 mm, planar laminated and ripple cross-stratified siltstone bases. The turbidites are separated by beds of laminated, dark grey to black mudstone (hemipelagite), which become thicker, and more common, upwards. At 550 m above the base, the ratio of hemipelagite to turbidite increases sharply to approximately 1:25, and, at 1200 m above the base, to about 1:1 (Figure 21); the latter defines the base of the Craig Hengae Member. Rare carbonate cone-in-cone concretions, up to 1 m in length and 80 mm thick, occur in the middle and upper part of the formation, mainly at the base of thin shelly sandstones. These are probably the 'rottenstones' of Pugh (1923) and Jehu (1926) and are most apparent in the north-east of the outcrop, although some occur as far west as Abergynolwyn [6718 0674].

Disturbed mudstones, 0.2–10 m thick, are recognised in most large exposures (Plate 8c) and are similar to those in equivalent strata around Llanymawddwy and Llanuwchllyn (Lockley, 1980a). Thicker disturbed beds, possibly up to 130 m thick, occur in the lower part of the formation at the type locality, although it is possible that these are composite (Figure 21). The thicker beds comprise massive, poorly cleaved, silty mudstones with wisps and 'rootless' folds of siltstone; thinner beds comprise slump folded, but coherent, mudstone and siltstone bands. Both types generally have planar tops and bases and can rarely be traced laterally for more than 1.5 km. At Nant Ceiswyn [7866 1340], a disturbed bed, 0.3 m thick, displays recumbent, isoclinal slump folds that are truncated by the overlying strata.

In the Cwm Amarch Syncline and north of Tal-y-llyn [7112 1051], a disturbed bed, 4–5 m thick, within thinly bedded siltstones and mudstones has a planar base and the lowest 50 mm are of massive pyritous mudstone with numerous matrix-supported sand grains. Above, is a zone of bedded silty mudstone with recumbent slump folds which become more upright towards the top of the bed and are truncated by the overlying turbiditic mudstones.

Sandstones are minor components of the formation and are mainly restricted to a 250 m-thick sequence, also with thin disturbed beds, 200 m beneath the Craig Hengae Member. The sequence, possibly the lateral equivalent of the Llaethnant Siltstones of the Llanuwchllyn dis-

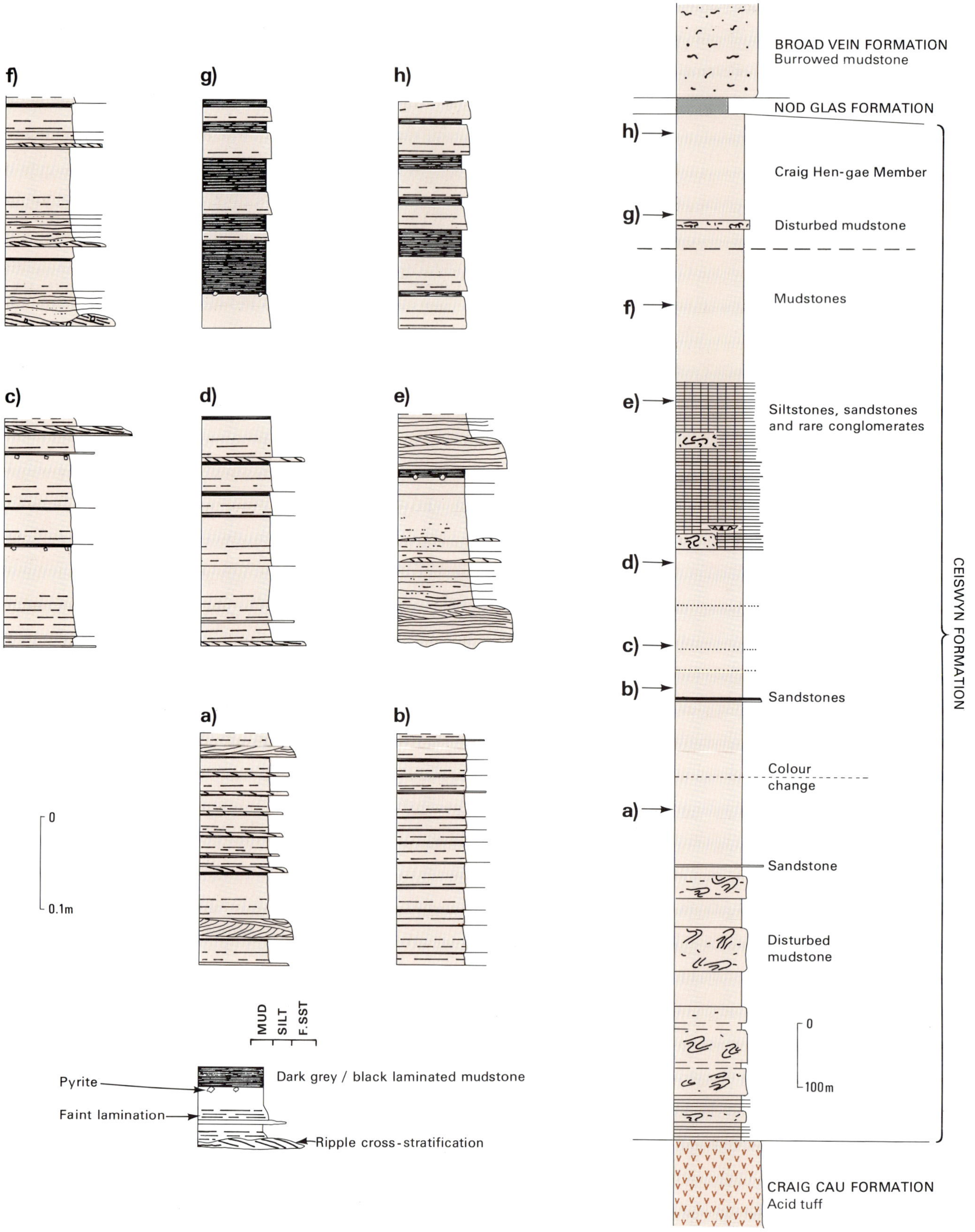

Figure 21 Composite section of the Ceiswyn Formation at Nant Ceiswyn [7876 1537] and Craig Hen-gae [7531 1076]. Detailed logs show typical small-scale sedimentary sequences.

trict (Lockley, 1980a), comprises graded sandstone-siltstone beds, 5–30 mm thick, interbedded with mudstones (Figure 21e). It is well exposed at Craig Hen-gae [7555 1048], north of Tal-y-llyn [7112 1051], and as far west as Abergynolwyn [6672 0660].

The formation also contains thin sequences of medium to thickly bedded sandstone up to 10 m thick. These occur mainly in the south-west of the district, in the lowest 300 m of the formation, and closely resemble similar sequences near the top of the underlying Ty'r Gawen Mudstone Formation. Near Bryncrug [6237 0377], a 10 m thick sequence comprises graded beds of medium- to fine-grained sandstone which thin and fine upwards from up to 0.7 m thick, at the base, to thinly interbedded, fine-grained sandstones and mudstones at the top. Most of the sandstone sequences are laterally impersistent, and can rarely be traced for more than 0.5 km. However, east of Bird Rock, a sequence maintains a constant thickness of 2.5–3 m for about 1.5 km, between [6518 0591] and [6537 0697].

In the east of the outcrop, thickly bedded sandstones are rare and only two sandstone sequences, each less than 2 m thick, are exposed in the 900 m of strata at the type section (Figure 21). Each comprises two or three closely spaced, normally graded, medium- to fine-grained quartzose sandstone beds up to 0.5 m thick. The upper sequence [7949 1470] occurs about 670 m above the base of the formation and contains thin (20–50 mm) lags of crinoid columnals, small brachiopods, starfish, graptolites and trilobite fragments. Palaeocurrent data imply transport from the north or north-north-west and the fauna indicates a shallow-water origin (Figure 22). In thin section, angular quartz grains in a matrix of diagenetic chlorite, illite and carbonate (E67011, E67016) can be distinguished. Common angular grains of amorphous chlorite and carbonate may be altered volcanic clasts (cf. Brenchley, 1970).

In the upper part of the formation, near Craig Hengae [7567 1106], a sequence some 8–9 m thick comprises two channelised conglomerate beds, up to 0.65 m

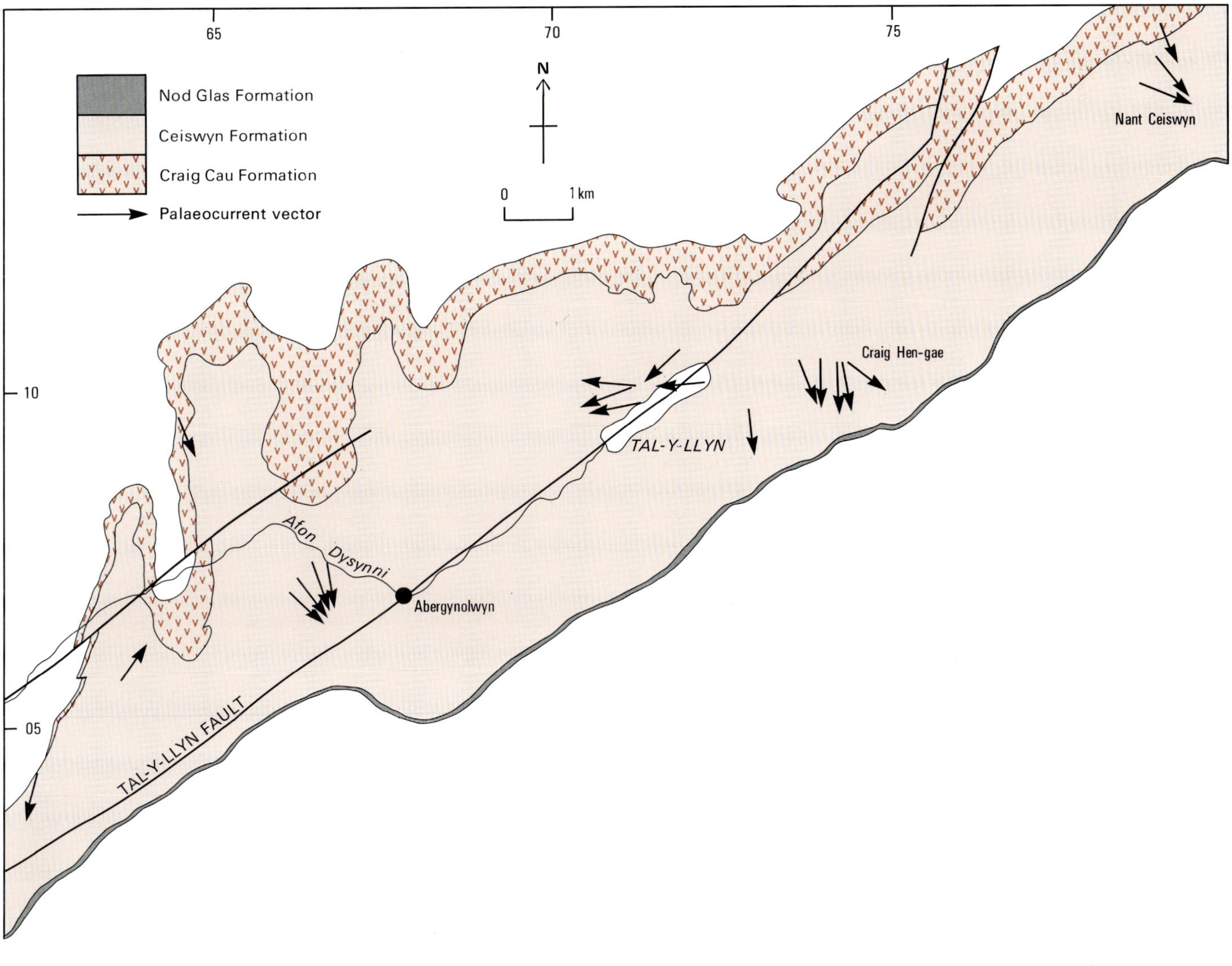

Figure 22 Palaeocurrent vectors from the Ceiswyn Formation (mainly determined from ripple cross-stratification).

thick, overlain by three normally graded sandstones, 0.1–0.4 m thick. The conglomerates are poorly sorted, with clast-supported basalt pebbles up to 50 mm long, and mudstone clasts. Many of the basalt clasts have been replaced by carbonate and are corroded by extreme pressure solution (E67114–5).

CRAIG HEN-GAE MEMBER

The Craig Hen-gae Member, 200–300 m thick, is characterised by an increased proportion of black, hemipelagic mudstone (Figure 21g, h). At the type section at Craig Hen-gae [7572 1026] it comprises thinly bedded black mudstones, up to 70 mm thick, and grey mudstone turbidites with rare, but disproportionately thick, basal siltstones. Pyrite cubes are common along the contact between the turbiditic and hemipelagic mudstones and within the rare siltstone beds. At least two beds of closely spaced black mudstones, each up to 3 m thick, occur in the lower part of the member. The member is also well exposed in crags south-west of Abergynolwyn [6626 0553] and in a synclinal outlier on the north side of the Tal-y-llyn Fault [6710 0670].

BIOSTRATIGRAPHY

Cox (1925, p. 566) reported the graptolites *Amplexograptus arctus* and '*Glyptograptus*' *euglyphus*, implying the *multidens* Biozone (Table 2), from near the base of the Ceiswyn Formation on Craig Cwm Ammarch. Near Bird Rock [6447 0564], sandstones from approximately 100 m above the base of the formation have yielded a shelly fauna with *Howellites* sp., *Sowerbyella* cf. *permixta*, *Broeggerolithus broeggeri*, *Flexicalymene* sp. and *Parabasilicus*? *powisii* (Plate 9d–g, i–k). The presence of *B. broeggeri* indicates that this fauna is early Soudleyan and is therefore younger than the Costonian or Harnagian fauna of the Derfel Member (Lynas, 1973), which immediately overlies the Aran Volcanic Group in the Bala district.

At the type section, transported fossils collected from 670 m above the base include discs of the ophiuroid *Protaster* cf. *salteri* and poorly preserved graptolites including *Climacograptus antiquus*. These are not very diagnostic of horizon, but *P. salteri* has been recorded from the lower Caradoc, and in the Bala district was collected by Elles (1922, p. 139) from Soudleyan strata.

East of the district, around Dinas Mawddwy, shelly faunas from within 50 m of the top of the formation include *Salterolithus caractaci* (Pugh, 1928, p. 351; BGS Zs 1071). The great thickness by which *S. caractaci* appears to overlie *Broeggerolithus broeggeri* is anomalous, because the recorded range of *S. caractaci* extends from Harnagian to Lower Soudleyan strata, but overlaps that of *B. broeggeri* only in the lower Soudleyan (Thomas et al., 1984). If the stratigraphical ranges of the trilobites are correctly assessed, it seems that nearly all of the Ceiswyn Formation must have been deposited during the early Soudleyan Stage.

CONDITIONS OF DEPOSITION

The upward increase of laminated black mudstones and decrease of the turbidite component within the formation probably reflects the gradual drowning of the shelf by eustatic sea-level rise (Dean, 1963; Leggett et al., 1981). Oxygen starvation occurred earlier within the district than in areas closer to the former shelf since the Craig Hen-gae Member, which was clearly deposited in anaerobic bottom waters, is equivalent to shallow-water sequences, deposited in oxygenated conditions farther east, for example Bala and Welshpool (Bassett et al., 1966; Cave and Price, 1978).

Palaeocurrent data, mainly from ripple cross-stratified siltstone and sandstone turbidites in the east of the outcrop, indicate transport from the north and north-east (Figure 22). No palaeocurrent data are available for the sandstones in the lower 300 m in the west of the outcrop. However, they are very similar to the north-westerly derived sandstones near the top of the Ty'r Gawen Mudstone Formation and probably represent continued influxes of small submarine fans.

All the disturbed beds are of typical Ceiswyn Formation strata and are interpreted to be entirely locally derived. Those beds with more coherent bedding display folds and thrusts, implying that they are the compressional, toe regions of slump sheets; the completely disaggregated beds, probably debris flow deposits, are interpreted as the distal equivalents. The concentration of disturbed beds in the lower part of the formation may reflect instability following the emplacement of the Craig Cau Formation. The thick accumulations between the Bala Fault and Bwlch Oerddrws may also reflect increased palaeo-slope, with shallowing towards the Bala district.

Nod Glas Formation

The Nod Glas Formation, of uppermost Caradoc age, comprises 20–30 m of intensely cleaved, locally graptolitic, black mudstone. The mudstones are soft and rusty or bleached by weathering because of a very high content of disseminated pyrite and, as a result, the outcrop forms a poorly exposed, but well-featured, depression. The calcareous and phosphatic facies of the formation, which typifies uplifted blocks within the Welsh Basin (Cave, 1965), is not developed and the facies is entirely that of the 'Corris Shale Member' (Lockley, 1980a). The type locality lies east of the district, at Nant y Nod [818 137] (Pugh, 1928; Lockley, 1980a). Within the district, the formation is best exposed at Ratgoed [7764 1171 to 7709 1145], in the north-east part of the outcrop (Figure 31). Precise thickness estimates are unreliable because of internal faults, but the outcrop width allows for only 15–20 m of strata. The base, seen near Foel Crochan [7709 1145], is gradational over 1.5 m and is marked by the disappearance of the mudstone turbidites of the Ceiswyn Formation.

The Nod Glas mudstones are mainly massive and uniform but, in places, a very faint lamination (0.25–1 mm), caused by a slight variation in grain size, can be determined. Locally, as south of Abergynolwyn [6767 0523], bedding is accentuated by concentrations of framboidal and cuboidal pyrite. In thin section, the main feature is the intense regional cleavage and large amounts of disseminated pyrite and carbonaceous material, locally concentrated in layers. Chlorite-mica stacks are common, although highly corroded by pressure solution, and analysis by X-ray diffraction indicates that the main component is chlorite; white mica occurs in only very small amounts (Roberts et al., 1991).

BIOSTRATIGRAPHY

Many graptolites, all characteristic of the *D. clingani* Biozone (Table 2), have been collected from the formation. Jehu (1926) and Pugh (1923) recorded *Dicranograptus clingani* itself. This suggests the lower part of the *clingani* Biozone because the species does not range higher than this zone in sections at Whitland in south-west Wales or Hartfell in Scotland. A cutting in the forestry track at Ratgoed [7764 1174] yielded *Climacograptus dorotheus, Dicellograptus morrisi* (Plate 9c), *Normalograptus mohawkensis, N. miserabilis* and *Plegmatograptus nebula. D. morrisi* indicates the upper part of the *clingani* Biozone. Other species recorded by Pugh (1923) include *Dicellograptus flexuosus, D. pumilus* and *Orthograptus* of the *calcaratus, quadrimucronatus* and *amplexicaulis* groups. Although some species recorded from the formation range into the *linearis* Biozone, none is definitive of that zone. From published records, it is possible that both the lower and upper parts of the *clingani* Biozone are present within the district.

East of the district, however, *D. clingani* is not recorded (Pugh 1928, p. 353) and part of the Nod Glas passes into impure shelly limestone (Pugh 1928, p. 355). Lockley (1980a, b) suggested that this limestone was equivalent to the Woolstonian (upper Longvillian) Cymerig Limestone of the Bala district, but, although he analysed the generic constitution of the fauna, he did not give any specific determinations in support of such an age. Around Welshpool, Cave (1965) showed that the lower part of the Nod Glas Formation includes a phosphorite that rests on upper Longvillian strata with possible non-sequence. There, the upper part includes *Dicellograptus morrisi* and an Onnian fauna including the trilobites *Onnia gracilis* and *Onnicalymene onniensis.* Therefore, the base of the graptolitic facies of the Nod Glas Formation appears to be diachronous, and, within the district, it appears to represent a greater proportion of the *clingani* Biozone than to the east. The base appears to rest on strata as old as Soudleyan, with no evidence of a non-sequence, and the upper part may be as young as Onnian.

CONDITIONS OF DEPOSITION

The formation represents the acme of oxygen starvation during the middle to late Caradoc (*D. clingani* Biozone) eustatic transgression (Dean, 1963; Cave, 1965; Leggett, 1980; Leggett et al., 1981). Sedimentation was entirely hemipelagic; the slight variations in grain size possibly reflect seasonal fluctuations. Stagnant bottom waters caused reduction of all Fe^{3+} ions to pyrite during early diagenesis, or even as aggregates developed upon organic debris on the sea floor (Curtis, 1977, 1980).

The palaeogeography of the district probably remained similar to that during Ceiswyn Formation times, with shallowing towards the north-east and north (Cave, 1965; Temple and Cave, 1992). Syndepositional faulting may have affected the distribution of the 'black shale' facies and Cave (1965) suggested that a horst in the Bala–Welshpool area confined coarse detritus to its eastern side and that the Nod Glas Formation was deposited on a sediment-starved slope.

Broad Vein Formation

The Broad Vein Formation, mainly bioturbated, pale grey, silty mudstone, is the lateral equivalent of the calcareous Moelfryn Mudstone Formation around Bala (Bassett et al., 1966) and the lower part of the Nant y Moch Formation of Aberystwyth (James and James, 1969; Cave and Hains, 1986). Pugh (1923) divided the Ashgill strata between the Nod Glas and Narrow Vein into the Broad and Red Veins, but as the latter, in places, is not distinguishable it has been abandoned (Martin et al., 1981; Leng, 1990). The base, exposed at Ratgoed [7764 1171], is marked by a change from black to pale grey mudstone over a thickness of 2–3 m. Although the base of the equivalent strata is unconformable around Bala (Bassett et al., 1966; Campbell, 1983), there is no sign of an unconformity within the Cadair Idris district.

Marked variations in thickness of the formation are probably due to excision or duplication of strata by faults within the Nod Glas Formation; the formation is only 200 m thick near Nant Cynfal [6260 0240], but less than 1 km to the south-west it is 400 m. It is about 550 m thick at the type locality, Foel Crochan [7670 1070], where the lower part comprises about 380 m of massive, bioturbated, silty mudstone and the upper part, 170 m of alternating massive silty mudstone and dark grey, locally graptolitic mudstone. The alternating lithologies of the upper part are well featured in outcrop (Plate 10a).

The main lithology is massive, pale grey, silty mudstone with branching, tubular burrows (*Chondrites*) infilled with darker, pyritous and more organic-rich mudstone. Bedding is defined locally by intensely bioturbated layers, ankerite cone-in-cone concretions up to 0.4 m long, and by sheets of early diagenetic phosphate concretions, 2–20 mm thick, which are themselves bioturbated. Rare mudstone turbidites, 0.03–0.1 m thick, with thin (0.5–2 mm) laminated siltstone bases occur where the bioturbation is less intense, for example at Briddellaw [7055 0664].

In the upper 170 m of the formation, the bioturbated mudstones are interbedded with sequences, up to 10 m thick, which are mainly of rusty-weathered graptolitic mudstone; these are the 'Red Vein' of Pugh (1923). In detail, the sequences comprise alternating beds, less than 10 mm thick, of dark grey, laminated (hemipelagic) mudstone and very thin, grey mudstone turbidites. The graptolitic mudstones are best exposed at Foel Crochan, Cynfal fach [6170 0150] and Dolgoch [6521 0435 and 6540 0427]. The top 20 m of the formation are well exposed in the underground slate workings beneath Foel Crochan; at least 8 m of dark grey laminated mudstone are overlain by 1.5 m of bioturbated silty mudstone with a layer of phosphate nodules, up to 30 mm in diameter, and an upper 14.5 m of massive, bioturbated silty mudstone to the base of the Narrow Vein Formation.

BIOSTRATIGRAPHY

The Broad Vein Formation yields a sparse shelly fauna, including the brachiopods *Christiania* and *Leangella*?

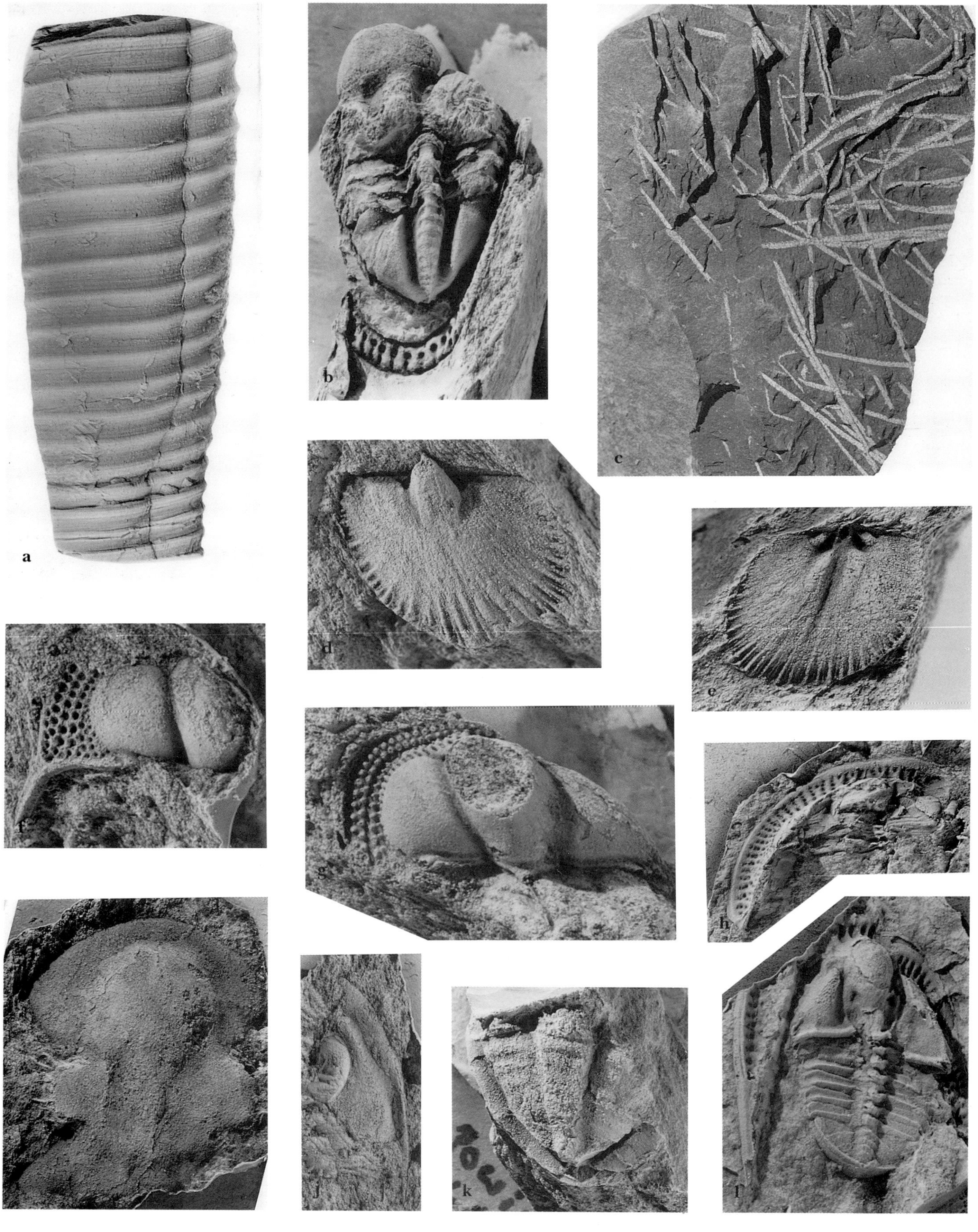
a
b
c
d
e
f
g
h
i
j
k
l

Plate 9 Examples of fossils from within and above the Aran Volcanic Group

a) Orthoconic(?) nautiloid cited by Pugh (1923, p. 523) as *Orthoceras perannulatum*. National Museum of Wales 27.110G.508, × 1. Narrow Vein Formation (Ashgill Series), Braich-goch Quarries, Corris [7486 0763]. The surface is marked both by annulations and by fine transverse and longitudinal threads; a similar sculpture is seen on upper Ashgill material from the Deganwy Mudstone, Conwy area, North Wales.
b) *Novaspis* cf. *albida* (Reed), fairly complete but disarranged exoskeleton, Zs 1050 (Pugh collection), × 3. Broad Vein Formation (Ashgill Series, Cautleyan Stage), Maes-y-gamfa Quarry [818 127].
c) Deformed *Dicellograptus morrisi* Hopkinson, *D. flexuosus* Lapworth and *Normalograptus mohawkensis* (Ruedemann), LF 115, × 1. Nod Glas Formation (Caradoc, *clingani* Biozone), west of Ty-cam, Cwm Ratgoed [7765 1170].
d), e) *Howellites* sp., internal moulds of pedicle valve LF44 and brachial valve LF 76, both × 4. Lower part of Ceiswyn Formation (Caradoc, Soudleyan Stage), west of Coed-y-gof, south of Bird Rock [6447 0564].
f), g) *Broeggerolithus broeggeri* (Bancroft), latex cast of LF 138, showing genal angle, and counterpart internal mould, LF 137, both × 4. Lower part of Ceiswyn Formation (Caradoc, Soudleyan Stage), west of Coed-y-gof, south of Bird Rock [6447 0564].
h), l) *Stapeleyella murchisonii* (Salter). h) latex cast of dorsal side of fringe RX 3271, × 2. l) latex cast of fairly complete exoskeleton, RX 3450, × 3. Cregennen Formation (Llanvirn Series, *artus* Biozone), south-west of Bryn Brith [6581 1486].
i), j), k) *Parabasilicus*? *powisii* (Murchison), latex casts of cranidium LF 74 and free cheek LF 40, and internal mould of pygidium LF 48, all × 2. Lower part of Ceiswyn Formation, Caradoc (Soudleyan Stage), west of Coed-y-gof, south of Bird Rock [6447 0564].

and several trilobites, including *Novaspis* cf. *albida* (Plate 9b), cyclopygids, *Opsimasaphus radiatus* and species of *Gravicalymene* and *Nankinolithus*; Price and Magor (1984) gave these a mid-Ashgill (Rawtheyan) age and regarded the '*Novaspis*-cyclopygid association' of the district as the deepest water assemblage of the various Rawtheyan trilobite associations found between Corris and the Bala district. Graptolite faunas from the 'Red Vein' are assigned to the *Dicellograptus anceps* Biozone (Table 2), probably its lower part. During this survey, *D. anceps*, *Orthograptus abbreviatus* and *O. fastigatus* were collected from near Aberllefenni [7636 0974]. The same species, together with *Normalograptus miserabilis*, were recorded at several other localities by Pugh (1923) and Jehu (1926). Neither the Nod Glas nor the Broad Vein formations give definite evidence for the presence of the *linearis* or *complanatus* biozones, nor for the corresponding shelly faunas of the Pusgillian and Cautleyan stages. However, it is possible that the faunal gap is represented by the unfossiliferous lowest part of the Broad Vein Formation.

Conditions of deposition

The formation marks the establishment of oxygenated conditions during early Ashgill times, probably due to shallowing; bottom waters and pore waters were sufficiently oxygenated to support burrowing activity in the cohesive mud substrate and cause diagenetic phosphate to crystallise at depths of only a few centimetres below the sea floor. However, concentration of pyrite within burrows implies localised 'micro'-reducing environments (e.g. Ekdale et al., 1984). The black mudstone beds in the upper part of the formation reflect sustained periods of oxygen and sediment starvation. The preservation of mudstone turbi-dites with siltstone bases suggests that turbidites were an important component, but the original proportion of hemipelagic mud is uncertain because of the thorough bioturbation. The preserved turbidites may represent temporary increases in sedimentation rate, which prevented the recolonisation of the substrate by burrowing organisms before successive turbidite deposition.

Narrow Vein Formation

The Narrow Vein Formation, which comprises 13–20 m of dark grey mudstones, is the lateral equivalent of the upper parts of the Moelfryn Mudstones of the Bala district (Bassett et al., 1966) and the Nant-y-Moch Formation of the Aberystwyth district (James and James, 1969; Cave and Hains, 1986). The mudstones are well cleaved and have been extensively quarried for slate. At the type locality, Bryn Eglwys Quarry, near Abergynolwyn [6942 0522], the base of the formation grades through 0.6 m from bioturbated pale grey mudstones of the Broad Vein Formation. The base is also well exposed in an air shaft on Foel Crochan [7706 1050] and in a collapsed cavern [7663 0985] in the Aberllefenni valley.

The formation is mainly homogeneous, dark grey mudstone, but a very fine lamination, with beds 1–4 mm thick, is defined locally by slight variations in colour and grain size. The darker layers are slightly richer in organic detritus and contain fewer silt grains. The lamination is most apparent in the middle of the formation and in the 2–3 m above the base and below the top. The lower mudstones weather with a brown tinge, possibly reflecting a small pyrite content. Rare *Chondrites*-type bioturbation has also been observed.

Biostratigraphy

The only macrofossils recorded from the formation are the undiagnostic orthocone '*Orthoceras perannulatum*' (Plate 9a) from Braich-goch Quarries [7486 0763], and a *Mucronaspis* (cited as '*Phacops*') sp. from Cymerau Quarry [7802 1153] (Pugh, 1923, p. 523). The microfossils, including palynomorphs and acritarchs (British Geological Survey Report WH/89/253R), are diverse, but have not been sufficient to assess biostratigraphically.

Conditions of deposition

The hemipelagic mudstones of the Narrow Vein Formation reflect quiescent basinal conditions. Preservation of fine lamination implies less oxygenated bottom waters than during deposition of the Broad Vein Formation. Leng (1990) considered that the strata were reworked by contour currents and reflected very low sedimentation rates, possibly as low as 5 mm/1000 years. Long wavelength

Plate 10a Excavations within the Narrow Vein Formation. Foel Crochan, Aberllefenni [7670 1065].

The craggy ground to the left comprises the lower part of the Broad Vein Formation, the central gullies are the laminated hemipelagic mudstones of the 'Red Vein' and the prominent ridge to the right represents the basal sandstones of the Garnedd-wen Formation.

(1 m), low amplitude (0.05–0.15 m) ripples, in light and dark mudstone laminae, described from a few kilometres east of the district (James and James, 1969; Leng, 1990), may support a contourite interpretation, but most of the formation is considered to represent hemipelagic fallout.

Drosgol and Brynglas formations

The Drosgol and Brynglas formations, comprising disturbed silty mudstones, sandstones and conglomerates (Jones, 1909; Cave and Hains, 1986) (Plate 10b), crop out only in Happy Valley and within the Ordovician inliers around Machynlleth. The formations are lithologically indistinguishable and can be separated only where the Pencerrigtewion Member, a mainly well-ordered sequence of sandstones and mudstones at the top of the Drosgol Formation, is recognised. However, in much of the district it is not recognised and the two formations are merged into the Garnedd-wen Formation (James, 1968, 1973).

Pencerrigtewion member

This member comprises up to 170 m of thinly bedded sandstone and mudstone with channelised and massive-bedded, quartzose sandstone and high-matrix sandstone. It is well exposed, about 20 m thick, near Forge [SN 7611

Plate 10b Sandstone pillows within disturbed silty mudstones. Drosgol Formation. Gelligoch [SN 7280 9970].

9977], adjacent to the Forge 'Overthrust', and the following sequence can be determined:

	Approximate thickness m
BRYNGLAS FORMATION	
Silty mudstone with wisps and clots of siltstone and quartzose sandstone. Very poorly cleaved (poorly exposed)	120
DROSGOL FORMATION	
Pencerrigtewion Member	
Sandstone, massive, fine-grained, quartzose with faint bedding	15–16
Mudstone, silty with numerous small clots of sandstone	0.1-0.2
Sandstone, massive, quartzose with faint lamination and rare mudstone flakes	0.9–1.1
Mudstone, silty with thin, ripple cross-stratified siltstone beds	0.6
Sandstone, massive, poorly graded, quartzose; generally structureless apart from faint convolute lamination in top 0.15 m	0.68–1.8
Mudstone, silty with thin ripple cross-stratified siltstones, particularly towards top; granular pyrite infills minor irregularities on upper surface	0.6
DROSGOL FORMATION (undivided)	
Mudstone, massive silty	more than 2

In Happy Valley, the member (about 170 m thick) is exposed in a quarry [SN 6213 9926]. The basal 20 m of quartzose sandstones, with interbedded mudstones, coarsen and thicken upwards. The sandstones are up to 0.35 m thick and the mudstone:sandstone ratio is approximately 2:1. Thinner sandstones commonly form starved and loaded ripples. Above, the sandstone content decreases over 50 m, with ripple cross-stratified beds, 10–20 mm thick, separated by layers of homogeneous silty mudstone, 0.02–0.2 m thick. A massive bed, about 7 m thick, of silty mudstone with contorted thin siltstone beds occurs near the top. East of the quarry [SN 6157

9966], the uppermost 100 m comprise disturbed massive silty mudstones with rafts of quartzose sandstone and sandstone balls up to 1 m in diameter.

Garnedd-wen Formation

The Garnedd-wen Formation, previously the Garnedd-wen Beds (Pugh, 1923; Jehu, 1926), occupies much high ground, including the Tarren hills [6850 0420], but is generally poorly exposed. Pugh (1923) described a type section south-west of Corris, on Mynydd Braich-goch [7392 0778] (Figure 23) and it is well exposed in Happy Valley [SN 6600 9960], Afon Cwmpandy [6335 0245], Braich-y-rhiw [6241 0146 to 6463 0008], Allt Gwyddgwion [6515 0123] and near Corris [7499 0786]. Because of widespread disruption of bedding, accurate thickness measurements are difficult, but it generally thins eastward, from about 1200 m in the south-west, to 445 m near Corris and 200 m at Aberllefenni (Figure 23).

The formation is mainly of massive, poorly cleaved, silty mudstone with isolated grains and pebbles up to 30 mm in diameter. Within the mudstones, anastomosing and conjugate surfaces, commonly slickensided, have been interpreted as syndepositional shear zones (Leng, 1990). Disaggregated, slump-folded and loaded, quartzose sandstone beds are typical of the formation. In many places, as in Happy Valley [SN 6240 9940], sandstone beds are disrupted into isolated sandstone balls, up to 1 m in diameter, within the silty mudstone.

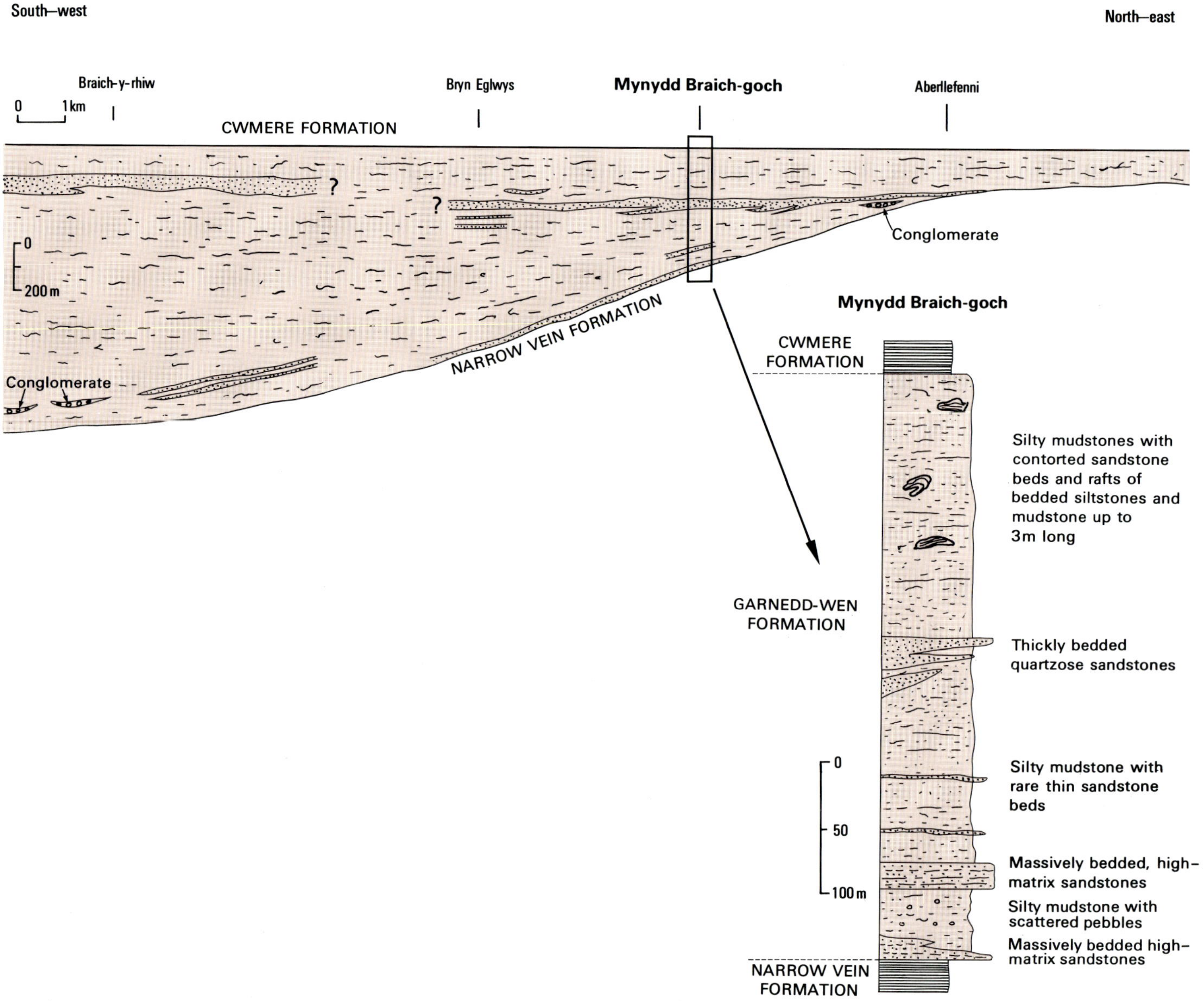

Figure 23 Fence diagram showing the broad lithological variations of the Garnedd-wen Formation and a simplified sedimentary log of the Mynydd Braich-goch type section [7392 0778].

Although mainly structureless, locally, as in Braich-y-rhiw valley [6254 0118], possible mudstone beds form ridges, some 5–15 m wide. Discrete rafts of bedded strata, up to 3 m in length, within the mudstones are common and some rafts may be much larger, possibly hundreds of metres in length and many metres thick. At Corris [7482 0809], a sequence of interbedded mudstone and sandstone, up to 22 m thick, which passes laterally into disturbed massive, sandy mudstone, is possibly a very large raft.

The base of the formation is well exposed in the numerous quarries in the Narrow Vein Formation. At Bryn-Egwlys quarries [6958 0535 and 6948 0529], the basal 20–30 m comprises poorly sorted, high-matrix sandstone, in ill-defined, massive beds up to 3 m thick, with thin interbedded silty mudstones. Some 5 km farther east, near Mynydd Braich-goch [7421 0863], these beds thin to 5 m and wedge out [7485 0762].

Sandstones

The formation includes two sequences of bedded, turbiditic, quartzose sandstone, which display marked lateral variations in thickness and bedding (Pugh, 1923; James, 1973; Leng, 1990). The lower sequence is confined to the west of the outcrop and the upper sequence, possibly equivalent to the Pencerrigtewion Member, progressively oversteps the underlying strata to the east until, at Aberllefenni, it overlies the Narrow Vein Formation (Figures 23 and 24). Comparable sandstone sequences are absent in the outcrop east of Aberllefenni.

The lower sequence comprises impersistent beds of quartzose sandstone, conglomerate and high-matrix sandstone, some of which are clearly infilled channels (Figure 23). Between Tywyn and Dolgoch [6500 0400], two sets of massively bedded sandstone, each up to 20 m thick, are separated by 20–50 m of sandy mudstone with scattered pebbles. The massive sandstone beds are normally graded, coarse to fine grained, with rare, planar-laminated tops. Many beds have irregular, erosional bases with flute marks up to 0.75 m long, commonly loaded, and others show convolute lamination and boudinage. Near Tywyn [6010 0049] (Jehu, 1926; James, 1973), a shallow channel is infilled with 15 m of crudely graded conglomerate beds. Each bed is up to 0.5 m thick, with clast-supported pebbles 0.01–0.1 m in diameter, in a fine-grained sandstone matrix. Vein quartz makes up more than 90 per

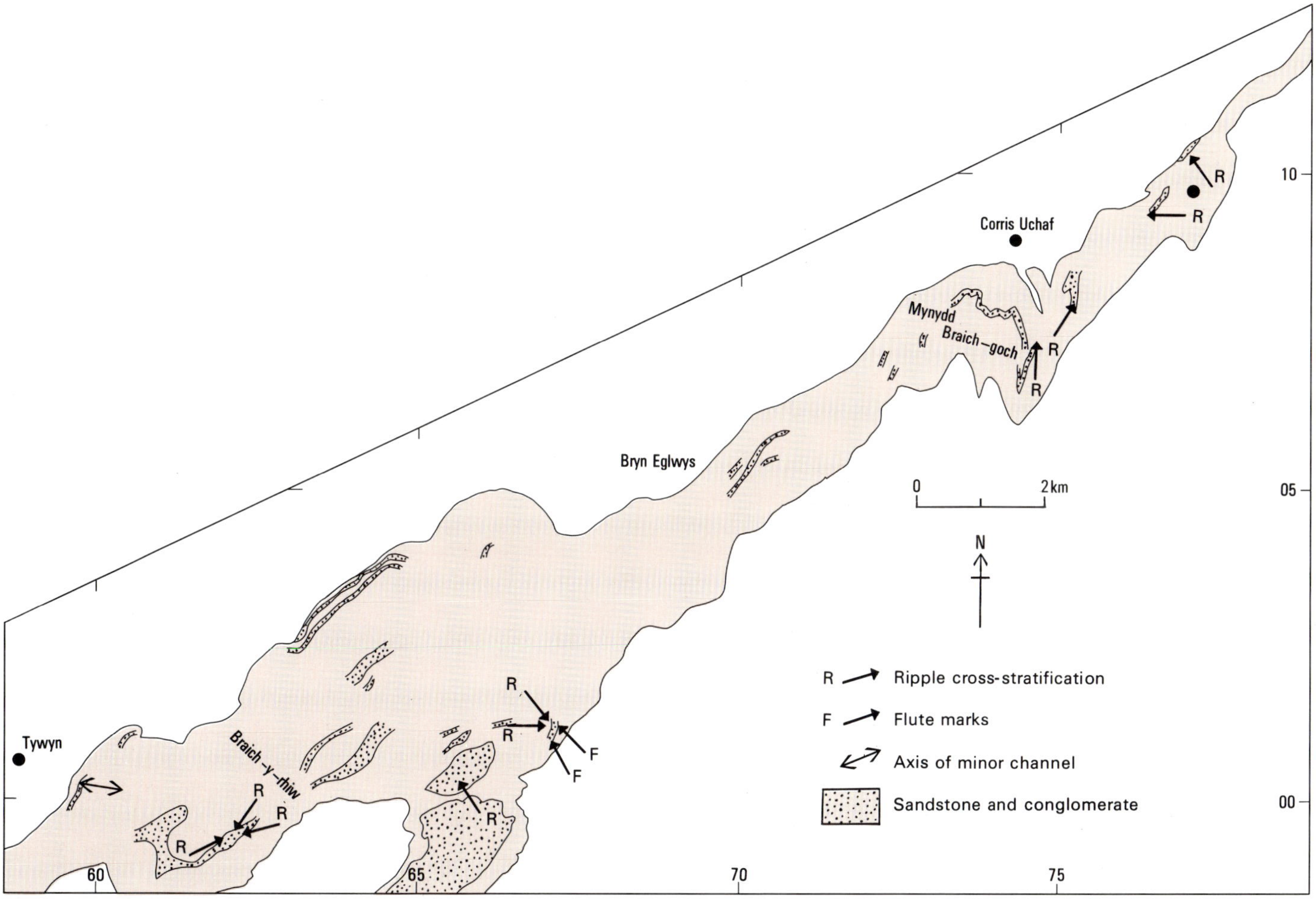

Figure 24 Distribution of sandstones and conglomerates and palaeocurrent data from the Garnedd-wen Formation. Each arrow represents the average of 6–10 readings.

cent of pebbles; other pebbles include sandstone, siltstone, mudstone, granite and acid tuff (Leng, 1990). Nearby [6080 0120], massively bedded quartzose sandstones pass laterally, over a few metres, into massive silty mudstone with isolated sand grains.

The upper sequence, the Llyn Barfog Member of James (1973), is also impersistent and crops out between the Braich-y-rhiw valley, in the south-west, and Aberllefenni (Figure 23). At Braich-y-rhiw [6322 0076] it comprises 80–90 m of quartzose sandstones, in beds 0.5–2 m thick, with interbedded mudstone, overlain by 20–30 m of thinner-bedded quartzose sandstones and mudstone. However, around Allt Nantgwenlli [6590 0045], less than 3 km to the east, it is 100 m of mainly massively bedded, quartzose sandstone. At this locality, ripple cross-stratification indicates transport towards the north-west, north-east and east (James, 1973). West of Corris, at the Mynydd Braich-goch type section, the upper sequence is up to 50 m thick and interdigitates with structureless silty mudstones (Figure 23).

East of Mynydd Braich-goch [7510 0795 and 7523 0825], 20–80 m of slumped sandstones, the likely correlative of the upper sequence, lie 150–200 m above the base of the formation. A few kilometres west of Aberllefenni [7666 0985], the same sequence, 20–30 m thick, lies 50–100 m above the base. Here, shallow channels of sandy mudstone crowded with small sandstone balls, less than 50 mm diameter, downcut into some sandstone beds and dewatering pipes transgress the contacts. In the vicinity [7660 0993], a lenticular bed of conglomerate, 10–15 m thick, occurs approximately 30 m above the Narrow Vein Formation. The massive conglomerate comprises normally graded, mainly vein quartz pebbles, up to 50 mm in diameter, with a small proportion of fine-grained sandstone matrix. At Aberllefenni [7707 1048], on the east side of the Afon Dulas, the equivalent of the upper sequence, 30 m thick, directly overlies the Narrow Vein Formation. The sandstones can be traced to the north side of Foel Crochan, but they wedge out within a few hundred metres and at Ratgoed [7860 1190] silty mudstones form the base of the formation.

High-matrix sandstones are also important components of the formation and occur in impersistent thin lenses (0.3 m long and 0.1 m thick) within silty mudstones, closely spaced massive beds (1–2 m thick) and rarely, for example in Allt Nantgwenlli [6631 0001], in massive, pebbly beds up to 10 m thick. The thicker beds, commonly composite, with undulating, erosional bases, are weakly, normally graded and locally, as near Ty'n-y-Ceunant [7433 0846], reverse graded. The sandstones commonly grade vertically and laterally into silty mudstones. Well exposed high-matrix sandstones occur south-east of Abergynolwyn [7050 0617] and in the cwm north of Tarrenhendre [6825 0438].

Sandstone petrography

The grains in the quartzose sandstones are locally well rounded, but more commonly their outlines are obscured by pressure solution suturing and quartz overgrowths. Quartz grains form up to 95 per cent of the rock, with accessory sodic plagioclase, muscovite and altered lithic clasts.

The sand fraction of the high-matrix sandstones is commonly bimodal, typically 0.05–0.1 mm and 0.25–0.5 mm across. Most grains are of angular quartz, commonly with authigenic overgrowths, with few of albite (5–10 per cent) Lithic fragments are mainly of silty mudstone, siltstone and sandstone, with less common basalt, acid tuff and fragments of phosphate nodules with enclosed ooliths.

Biostratigraphy

No macrofossils have been recorded within the district, but to the east, near Dinas Mawddwy, *Mucronaspis mucronata*, which ranges from topmost Rawtheyan to Hirnantian, occurs in the lower part (Pugh, 1928). Diverse microfossil assemblages have been recovered, but they do not distinguish an age more precise than late Ordovician.

Conditions of deposition

The Garnedd-wen, Drosgol and Brynglas formations were deposited during the Ashgill (Hirnantian) glacioeustatic regression, when sediment input from the shelf increased (Cave, 1979; Brenchley, 1984; Brenchley and Newall, 1984; Cave and Hains, 1986). Leng (1990) proposed that the formation comprises sedimentary mélanges and that the sandstone sequences developed at the lowest sea-level stands during the regression.

The massively bedded sandstones, and the Pencerrigtewion Member, are interpreted as proximal turbidites, mainly channel fills, and the thinly bedded sequences as overbank deposits. The high-matrix sandstones reflect higher concentration flows with some of the thicker beds being debris flow deposits. Palaeocurrent data are diverse (Figure 24), perhaps reflecting a situation at the base of a slope, but the north-westerly thinning of the sandstones suggest shallower water to the east and north-east of the district (James and James, 1969; Cave and Hains, 1986). The progressive eastward overstepping of many hundreds of metres of strata by the upper sandstone sequence suggests a period of non-deposition (cf. James, 1985), or erosion, in the outcrop beyond Corris.

The sandstones and conglomerates are mainly of quartz and sedimentary rock clasts, implying little input from igneous or metamorphic sources. The rounding of pebbles and sand grains implies shallow-water working prior to their incorporation into turbidites. However, the diverse mineralogies and angular grains of the high-matrix sandstones suggests little shallow-water reworking.

Slumping and liquefaction of the substrate was initiated by emplacement of thick disturbed beds or turbidites. The conjugate shear zones within the silty mudstones were probably caused by sudden loading and dewatering (Leng, 1990). Many of the isolated sandstone balls, which typify the formation, probably developed by extreme loading of sandstone beds, but there is also abundant evidence of lateral, downslope boudinage of bedded strata.

FOUR

Silurian

Silurian strata, up to 1.5 km thick and entirely of Llandovery age, succeed the Ashgill strata conformably. Early work on the Silurian sequence within, and about the district was dominated by Jones (1909), Jones and Pugh (1916, 1935) and Pugh (1929), with other contributions by Jehu (1926), Cave (1979) and Martin et al. (1981). More recently, Cave and Hains (1986) modified the lithostratigraphical subdivisions of these earlier workers and formally defined the type localities of many of the formations; it is mainly their stratigraphy that is used in this account (Figure 25). The Ashgill/Llandovery boundary lies within the lower part of the Cwmere Formation and for convenience the uppermost Ordovician strata are described here.

The Llandovery formations occupy about 80 km^2 of the south-east corner of the district, between Penegoes, Pennal and Corris. The outcrops of the Cwmere, Derwenlas and Cwmsymlog formations form a narrow, north-east–south-west-striking band between the Happy Valley, in the south-west, and the east margin of the district. Minor outcrops also occur in the Forge and Gelligoch inliers, to the west and east of Machynlleth. The outcrop of the thicker Devil's Bridge, Borth Mudstones and Blaen Myherin Mudstones formations is much larger and partly due to the broadening caused by the reduction in fold plunge in the south-east of the district.

Much of the Llandovery outcrop is concealed by dense conifer plantations, but the forestry roads provide both useful access and exposures. The strata form poorly exposed, but well-featured terrain, mainly between 200 m and 300 m above OD. Scarps reflect sandstone-rich horizons, and hollows mark the position of the mudstones. Typical well-featured ground occurs south of Corris [7630 0530], defining a major periclinal anticline, and north of Machynlleth [7580 0260].

The Llandovery formations are mainly distinguished by the thickness and sandstone content of the turbidite rhythms and the amount, and type, of hemipelagic mudstone (Cave, 1979). The hemipelagites, are considered to represent 'background' sedimentation and comprise both black/dark grey laminated mudstone, deposited in anaerobic bottom waters, and pale grey, bioturbated mudstone, deposited under oxygenated conditions. Phosphate nodules are widespread in the bioturbated mudstones and probably mark the contact between oxygenated and reduced pore waters during very early diagenesis (Cave, 1979; Smith, 1987b). Partially burrowed, dark grey, mudstones, in which lamination is preserved, imply intermediate, poorly oxygenated conditions. Sandstones and siltstones are thin or absent in the Rhuddanian and Aeronian sequence, but are more prominent in the Telychian sequence (Figure 25).

The strata were deposited during the eustatic transgression that followed the uppermost Ashgill (Hirnantian) glacioeustatic regression (Brenchley, 1984; Brenchley and Newall, 1984). By comparison with the Ordovician sedimentary sequence, there is little evidence of syndepositional disturbance. However, it is considered that the form of the Welsh Basin remained similar to that established during Ordovician times, with the slope from the basin to the shelf corresponding to the Welsh Borderland Fault System (Ziegler et al., 1968; Cope et al., 1992). In the Cadair Idris district the Llandovery strata are free of non-sequences and unconformities, and sediments were transported into the district from several directions.

The Llandovery strata record two distinct phases of basin evolution. The Cwmere, Derwenlas and Cwmsymlog formations (Figures 25 and 26), which are closely similar to the succession in the Aberystwyth district, maintain similar lithological characteristics across the district. They are thicker, approximately 180 m, than the equivalent strata in north-west Wales, where the *acinaces* to *turriculatus* Biozones are condensed into only 32 m of strata (Baker, 1981), and the sequence indicates relatively low sedimentation rates and displays limited diachroneity.

By contrast with the Rhuddanian and Aeronian strata, the Telychian Devil's Bridge and Borth Mudstones formations interdigitate and were deposited during a relatively small part of the *turriculatus* Biozone (*renaudi/utilis* subzones) (Figures 25 and 27). They marked a new phase of basin development during which subsidence and sedimentation rates increased and syndepositional faults became more important in constraining turbidite deposition. Because the Telychian formations record the complex interplay of turbidite fans derived from different quadrants, boundaries between formations are commonly based on subtle changes in the ratio of sandstone to mudstone.

The Silurian lithostratigraphy is constrained by highly refined biostratigraphical control based on graptolite biozones (Figure 25). Within the district, recent collections, and those of Jones and Pugh (1916), suggest that all the graptolite zones from the uppermost Ashgill *persculptus* Biozone to the lower Telychian *turriculatus* Biozone are present. Within the district graptolites are rare in the Telychian strata, but elsewhere within the sequence they are moderately abundant, especially within the dark grey, laminated, hemipelagic mudstones where they are preserved as pyrite internal moulds of low to full relief. In well-cleaved strata their identification is difficult.

The Silurian graptolite zones are, on average, less than a million years long (Rickards, 1976, 1989; Zalasiewicz, 1990). However, with the use of formal and informal subdivisions, time segments of less than 500 000 years may be recognised. The zonal system used is that of Rickards (1976), and modified, with subdivision of the *turriculatus*

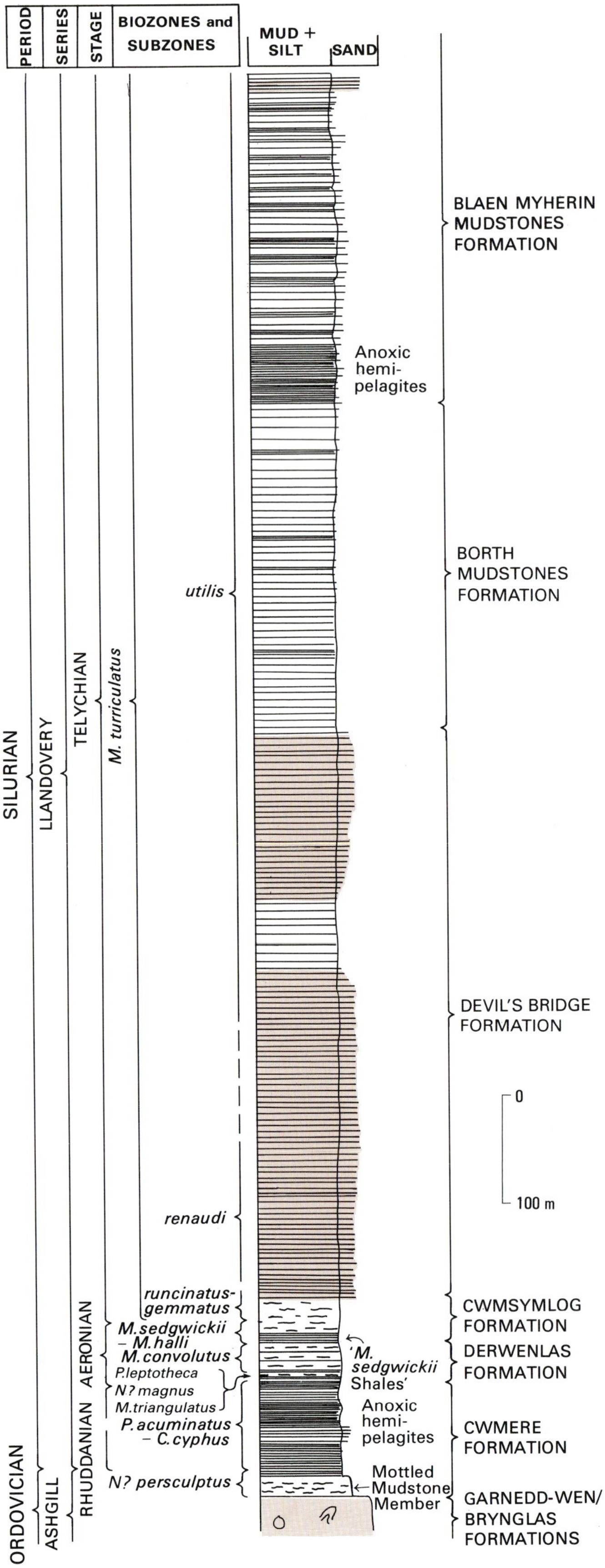

Figure 25 Stratigraphical column of the Silurian sequence of the Cadair Idris district.

Biozone, by Loydell (1991). The stratigraphical position of the Telychian graptolite collections are also indicated in Figure 27. The main characteristics of each biozone and subzone are given in Appendix 1, important graptolite localities in Appendix 2, and a range chart of selected species is given in Table 3. Details of other graptolite localities are given in the appropriate Open File Reports (p. viii).

Cwmere Formation

This formation is equivalent to the Cwmere Group of Jones and Pugh (1916) (Cave and Hains, 1986). With the exception of the Mottled Mudstone Member at its base, the formation is mainly thinly bedded, dark grey laminated (hemipelagic) mudstone with thin mudstone turbidites, in pairs some 0.04–0.1 m thick. The Mottled Mudstone Member, and a few metres of the overlying strata, lie within the *persculptus* Biozone (uppermost Ashgill, Ordovician) (Figure 25), formerly considered to be of lowest Llandovery age (Cocks et al., 1984). As in the Aberystwyth district (Cave and Hains, 1986), only the *acuminatus* to *cyphus* Biozones have been proved in the remainder of the formation. However, faunas have not been retrieved from the upper part and it is possible that it extends into the *triangulatus* Biozone.

The formation crops out within two inliers near Machynlleth, but the main outcrop lies between Mynydd y Llyn [SN 6658 9888], in the south-west, and Mynydd Hendre-ddu [8005 1210], in the east. The soft, fissile mudstones form a depression but, in areas of low dip, as at Mynydd Rhyd-galed [7010 0410], small ridges correspond to slightly sandy strata (Plate 11a). In many places the base of the formation is marked by a prominent dip slope on the top of the Garnedd-wen or Brynglas formations.

The Cwmere Formation is mainly about 100 m thick, but it is markedly attenuated in fold limbs and thickened in fold hinges. Well-exposed sections occur at Cwmffernol [6715 0040], Mynydd Rhyd-galed, Cwm Cadian [7423 0625, 7290 0655] and north-east of Cymerau Farm [7772 1077 to 7847 1129].

Mottled Mudstone Member

This member, the lowest 10–25 m of the Cwmere Formation, sharply overlies the phacoidally cleaved, silty mudstones of the Garnedd-wen or Brynglas formations. It is more compact and less prone to weathering than the remainder of the formation and is best exposed at Cwmffernol [6678 0067], and, in the north-east of the district, at Cwm Cadian [7442 0612 and 7285 0663]. It comprises alternations of medium grey mudstone turbidites, 0.05–0.3 m thick, commonly with thin siltstone bases, and pale grey, *Chondrites*-burrowed, rarely laminated, mudstones, 0.02–0.1 m thick. The latter are probably mainly hemipelagic, but may also include the upper, oxidised portions of turbidites. Small phosphate nodules are commonly developed along the contact between the pale and dark mudstones. Bioturbation was insufficient to destroy the bedding and most burrows are small, less than 2 mm in diameter. Widely spaced, bedding-parallel sheets of pyrite up to 10 mm thick occur at the base of some turbidites. A layer, 0.3 m thick, of dark grey, laminated

Table 3 Ranges of selected graptolite species through the uppermost Ordovician *(persculptus)* and Llandovery biozones and subzones of the Cadair Idris district.

Biozones	persculptus acuminatus	atavus acinaces	cyphus	triangul-atus	magnus	leptotheca	convolutus	sedgwickii	turriculatus (renaudi)	turriculatus (utilis)
N ? persculptus	×									
N ? parvulus	×									
N. normalis	×									
N. angustus	×	×								
G ? avitus	cf.	×								
A. atavus				×						
Rh. toernquisti		×	×	×	×	×				
N. medius		×								
Rh. extenuatus		×								
A. gracilis		×								
N. rectangularis		?								
Me. hughesi		×								
M. austerus s.l.		?	×	×						
M. difformis			×							
L. acinaces			?							
Co. cyphus			?							
'O'. mutabilis			cf.							
Me. undulatus			cf.	×	×		×	×		
M. triangulatus separatus				×						
M. triangulatus triangulatus				×						
'O'. insectiformis				cf.						
'G'. sinuatus sinuatus				×						
G. aff. *incertus*				×						
M. communis communis				×						
P. concinnus				×	×					
N ? magnus					×					
M. triangulatus fimbriatus					×					
M. chrysalis					?					
M. pseudoplanus					×					
Pr. leptotheca						×	×			
M. argenteus						×				
M. denticulatus						×				
Mcl. sp.						×				
Co. gregarius						×				
G. tamariscus s.l.						×				
M. convolutus							×			
M. lobiferus							×			
M. limatulus							×			
Ra. hybridus hybridus							×			
Cl. retroversus							cf.			
N ? scalaris							×			
Ra. spina							×			cf.
P. regularis							cf.			
M. decipiens							cf.			
M. sedgwickii								×		
La. tenuis								×		
M. contortus								×		
Gl. incertus								×		
S. pseudoruncinatus									×	
S. plumosus									×	
P. renaudi									×	×
M. bjerreskovae									×	
Ra. maximus									×	×
M. halli									×	
M. utilis									?	×
M. turriculatus										×
M. planus										×
M. marri										×
M. involutus										×
P. bjerringus										?
Pe. conicus										×
Pe.giganteus										cf.
S. barrandei										×
Pe. tenuis s.l.										×
P. nudus s.l.										×

hemipelagic mudstone with *Normalograptus? persculptus*, the '*persculptus* band' of Jones and Pugh (1916), occurs approximately 1 m above the base of the formation.

Although the member is dominantly of mudstone, it locally includes discontinuous sandstones. For example, west of Mynydd Rhyd-galed [7000 0418], well-bedded mudstones include sandstone clots, probably load balls, 0.1–0.15 m in diameter. Near Machynlleth [7265 0059] the member contains parallel-laminated and ripple cross-stratified sandstones up to 10 mm thick.

Strata above the Mottled Mudstone Member

Because of a high content of disseminated pyrite, bedding and cleavage planes in these strata are rusty weathered and stained with ochreous material (Plate 11b). The mudstone turbidites are either massive or faintly laminated, and some contain normally graded siltstones and fine-grained sandstones, 5–10 mm thick, at the base. These generally have sharp bases or minor load structures, but rarely display internal lamination. A sequence of sandstone-rich turbidites, 3–4 m thick, in the middle of the formation around Derwenlas, in the south of the district, is similar to that determined in the Aberystwyth district (Jones and Pugh, 1916; Pugh, 1928; Cave and Hains, 1986).

The hemipelagic mudstones are rich in pyritised graptolites, fine-grained organic debris, disseminated pyrite and small, ellipsoidal monazite nodules, about 1 mm in diameter (Milodowski and Zalasiewicz, 1991b). Individual carbonaceous laminae, less than 1 mm apart, are possibly of algal origin (cf. Rickards, 1964) and chlorite-coated, folded graptolite rhabdosomes are common. Layers of disseminated pyrite, generally less than 2 mm thick, but up to 30–40 mm, locally form up to 10 per cent of the mudstone. In the upper part of the formation, pyrite nodules are common, both isolated and in beds up to 0.1 m thick. In Cwm Cadian [7287 0660], the disc-shaped nodules, up to 60 mm in diameter, are overgrown by radiating aggregates of quartz and white mica. Locally, as near Aberllefenni [7777 1083], the mudstones contain layers of disc-shaped siliceous mudstone concretions with disseminated pyrite. The concretions are up to 70 mm in diameter and parallel to bedding, which is compacted around them.

In places, rare isolated clasts indent the mudstone substrate and are draped by succeeding hemipelagites. East of Pennal [7182 0033], a subrounded clast of feldspathic greywacke, with dimensions of $0.1 \times 0.05 \times 0.05$ m, occurs within graptolitic mudstone. Although it might be a glacial dropstone, graptolite collections indicate that the enclosing strata are much younger (upper *atavus–cyphus* biozones) than the Hirnantian glaciation. Instead, the clast may have been transported by an algal mat or algal holdfast.

Conditions of deposition

The Mottled Mudstone Member is the first indication of the transgression which followed the Hirnantian regression (Brenchley, 1984; Cope et al., 1992). It was deposited mainly under oxygenated conditions, but the thin interbeds of black mudstone, in particular the *persculptus* band, record anaerobic conditions. Sediment input into the district declined in late *persculptus* times and thereafter the bottom water was anaerobic. Hemipelagic sedimentation dominated, with few incursions of dilute turbidity currents which deposited thin mudstones. The similarity of the formation with that in the Aberystwyth district, particularly the central sandstone-rich sequence, suggests a similar derivation, possibly from the south-east (Cave and Hains, 1986).

Derwenlas Formation

This formation spans the *triangulatus* to *convolutus* Biozones of the Aeronian (Figures 25 and 26) and comprises mainly thinly bedded turbiditic and bioturbated, hemipelagic mudstones, with minor siltstones and fine-grained sandstones. It is broadly equivalent to the lower part of the Derwen Group of Jones and Pugh (1916) (Cave and Hains, 1986). The formation forms a distinctive, mainly well-exposed scarp (Plate 11a) and its base is exposed at Cwmffernol [6730 0030] and Mynydd Rhyd-galed [7010 0412], in the centre of the district. At Derwenlas, the type locality [SN 7187 9915], near the southern edge of the district, the formation is 37 m thick and in the Forge Inlier [7590 0130] it is 25–30 m thick. Between Cwmffernol and Cwm Cadian [7343 0667] it is estimated to be 40–45 m thick.

At the Derwenlas Formation type section (Jones and Pugh, 1916, pp. 355–356) six beds, 30 mm to 2 m thick, of dark grey, graptolitic mudstone with only thin turbiditic mudstone interbeds can be distinguished (Figure 26). These beds were named after their most characteristic graptolite and were correlated with a similar succession in the Rheidol Gorge [SN 7590 7970] of the Aberystwyth district (Jones, 1909). Jones and Pugh (1916) placed the base of the formation at the '*Magnus*' band, a layer, 25 mm thick, of black mudstone with abundant specimens of *Mesograptus* (now *Normalograptus?*) *magnus* (*magnus* Biozone). However, the base is now revised to 5 m beneath the '*Magnus*' band, at the bottom of a transition zone with the Cwmere Formation, probably of *triangulatus* Biozone age (Figure 26). In this zone, there are fewer laminated hemipelagite mudstones and more pale turbiditic, and bioturbated (oxic), hemipelagic, mudstone beds. The base is also well exposed west of Mynydd Rhyd-galed, where turbidites at the top of the Cwmere Formation are also much siltier.

The bulk of the formation is thinly bedded, but it commonly appears as massive and poorly bedded and has an irregularly developed cleavage. Bedding is better defined near the top of the sequence. Massive and faintly laminated, turbiditic mudstones are the dominant lithology, but thin (less than 10 mm) pale grey hemipelagic mudstones and varying proportions of siltstone, with ripple cross-stratification, planar lamination and mudstone rip-up clasts, also occur. The turbidite/hemipelagite pairs are generally less than 50 mm thick and the basal sandstone layers are less than 10 mm thick. Phosphate concretions are common and some of the hemipelagic mudstones display *Chondrites*-type bioturbation. In the south-west of the district, at Cwmffernol, thicker, 20–30 mm, siltstones are developed within

Plate 11a Featuring within the gently dipping Cwmere Formation at Mynydd Rhyd-galed [7015 0410]. The prominent gully approximates to the base of the Derwenlas Formation.

Plate 11b Typical Cwmere Formation strata. Bedding dips gently to the right (south-east) and cleavage dips steeply to the right. Mynydd Rhyd-galed [7010 0411].

strata of the *convolutus* Biozone and, generally, the overlying turbiditic mudstones are also thicker (0.05–0.1 m).

Conditions of deposition

The formation reflects a period of higher sedimentation rate than the Cwmere Formation and deposition in mainly oxygenated bottom waters. The thin graptolitic beds represent temporary establishment of the anaerobic conditions which typified the Cwmere Formation. The direction of sediment input is uncertain, but near Llanymawddwy, only 10 km to the north-east of the district, the formation is reduced to 14 m, is calcareous and contains shelly fossils (Pugh, 1929), suggesting deposition in a shallower-water environment. The implied proximity of a shelf to the east or north-east suggests that turbidity currents may have been derived from this direction.

Cwmsymlog Formation

This formation comprises thinly bedded mudstone turbidites, 0.03–0.1 m thick, with basal siltstone/sandstone layers rarely thicker than 20 mm, interbedded with thin pale grey, bioturbated hemipelagic mudstones (Figure 25). Formerly the upper part of the Derwen Group of Jones and Pugh (1916), it was defined by Cave and Hains (1986) as the strata above, and including, the '*Monograptus sedgwickii* Shales' (*sedgwickii* Biozone) of Jones and Pugh (1916), but below the Devil's Bridge Formation.

Plate 11c Sandstones within the Devil's Bridge Formation. Cefn [7208 0036].

The presence of *renaudi/utilis* (mid-*turriculatus* Biozone) faunas from near the base of the overlying Devil's Bridge Formation suggests that the Cwmsymlog Formation includes the *runcinatus* and *gemmatus* subzones of the *turriculatus* Biozone (cf. Loydell, 1991) (Figure 25). The formation is probably partly equivalent to the Rhayader Mudstones Formation of the Rhayader/Llanilar district (Davies et al., in press).

The formation is mainly 20–40 m thick. At the north end of the Forge Inlier [7580 0125], the formation is about 37 m thick, including 2–5 m of '*M. sedgwickii* Shales'. It is best exposed in Cwm Cadian [7343 0667] where it is 40 m thick. At the Derwenlas type locality [SN 7190 9910], the formation is 24 m thick and includes the thickest sequence, 6 m, of the '*M. sedgwickii* Shales' in the district (Figure 26). Elsewhere, good sections occur at Cwmffernol [6725 0032], near Aberllefenni [7791 1065, 7810 1091] and at the east margin of the district [7939 1203].

Lithologically, the formation contains more sandstones than the Derwenlas Formation, but less than the Devil's Bridge Formation. The strata are better cleaved and bedding is more clearly defined than in the Derwenlas Formation. The base is sharply defined by the '*M. sedgwickii* Shales', which form a depression at outcrop, for example at Ffridd Coed [7791 1068] and at Garth [7589 0141]. However, it is difficult to locate the base in areas where the member is poorly exposed, for instance at the north end of the Gelli-goch Inlier [7161 0003].

'M. SEDGWICKII SHALES'

The member, 2–6 m thick, is equivalent to the 'Dark grey mudstone member' of the Aberystwyth district (Cave and Hains, 1986). It closely resembles lithologies in the Cwmere Formation and comprises pairs, 6–80 mm thick, of normally graded and faintly laminated turbiditic mudstone beds overlain by dark grey, laminated hemipelagic mudstone beds. In places, as east of Mynydd Hendre-ddu [7936 1205], thin siltstone beds impart a distinctive striped appearance to weathered surfaces. Thin seams of pyrite, up to 2 mm thick, are common and the hemipelagic mudstones contain abundant *M. sedgwickii*.

STRATA ABOVE THE 'M. SEDGWICKII SHALES'

These strata, which span the upper *sedgwickii* to lower *turriculatus* Biozones, are mainly grey and pale grey, turbiditic and hemipelagic mudstones with basal sandstone layers up to 20 mm thick. Each turbidite/hemipelagite pair is generally less than 50 mm thick. Thin beds of laminated hemipelagic mudstone, a short distance above the '*M.sedgwickii* Shales' (Figure 26), were assigned to the *halli* Biozone by Jones and Pugh (1916) (see also Loydell, 1991).

The strata are distinctively colour banded, in varying shades of brown, but the 'green mudstone member' of the Aberystwyth district (Cave and Hains, 1986) is present only on the southern edge of the district, south of Forge [SN 7640 9888]. At Aberllefenni [7805 1070], typical strata comprise basal sandstone layers, 5 mm thick, overlain by 20–30 mm of turbiditic mudstone and 10 mm of pale grey, hemipelagic mudstone with *Chondrites* burrowing and phosphate nodules. Soft sediment deformation structures are widespread and at Happy Valley [SN 6430 9961] and east of Corris [7592 0773], these include low-angle, syndepositional extensional faults, small-scale slump folds, loaded sandstone beds and flame structures.

CONDITIONS OF DEPOSITION

The '*M. sedgwickii* Shales' were deposited during a transgression (Ziegler et al., 1968; Cope et al., 1992). The flooding of the shelf reduced the input of detritus into the district and there was widespread oxygen starvation in the basin. The overlying strata accumulated mainly by turbidity current deposition and by hemipelagic settling, under oxygenated conditions. The form of the basin was

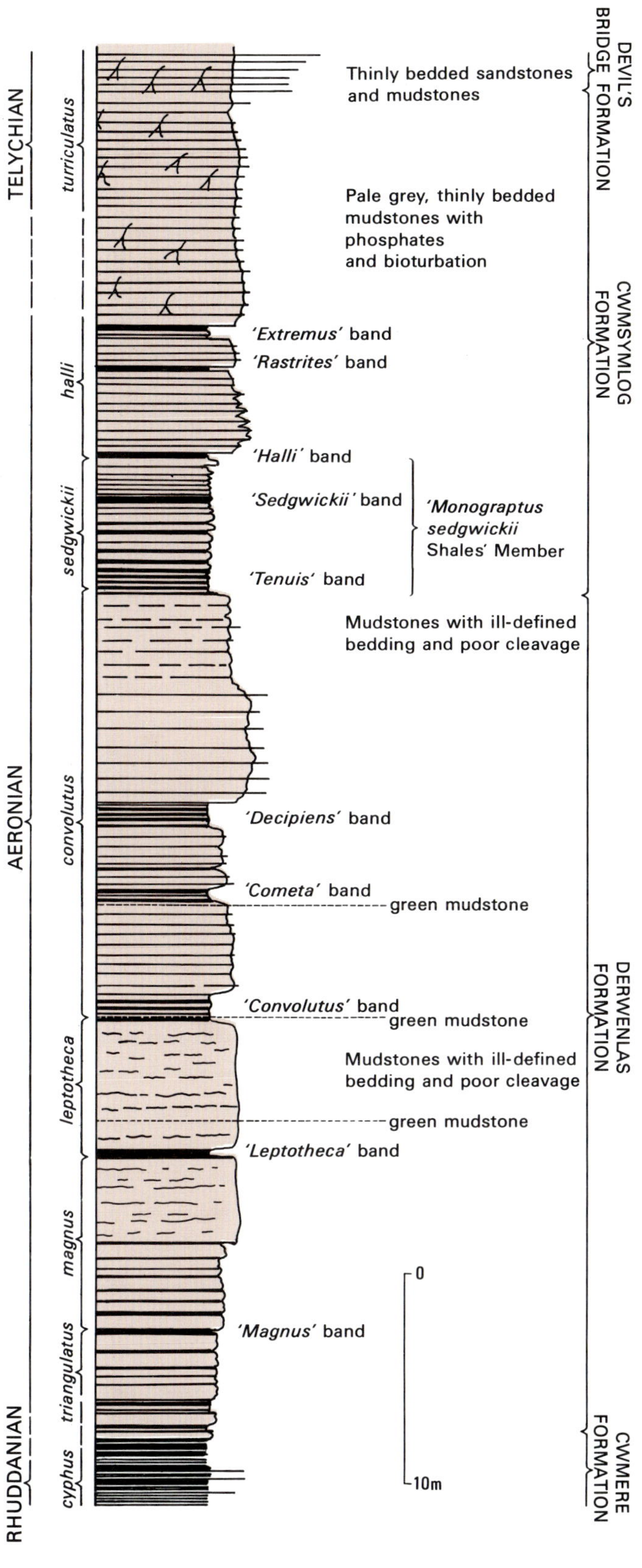

Figure 26 Simplified log of the Derwenlas and Cwmsymlog formations at the Derwenlas type locality [SN 7187 9915] showing the distribution of dark grey, laminated, hemipelagic mudstone (modified from Jones and Pugh, 1916).

probably similar to that during Rhuddanian and Aeronian times. The lack of internal bedforms makes determination of the provenance of the turbidites difficult. Cave and Hains (1986) considered they were derived from the south-east, but derivation from the Bala and Berwyn districts, to the east and north-east, is possible as there is evidence there of broadly contemporaneous nondeposition or erosion (Temple, 1988; Cave, in Cope et al., 1992).

Devil's Bridge Formation

This formation, defined by Cave and Hains (1986), is a modification of the Devil's Bridge Group of Jones (1909) and lies within the *renaudi* and *utilis* subzones of the *turriculatus* Biozone (Figure 27). It crops out extensively in the south-east of the district, forming most of the Llandovery outcrop, and comprises variable proportions of interbedded turbiditic sandstone and mudstone with thin, mainly 'oxic' hemipelagic mudstones. Forestry tracks north and north-east of Pennal [7336 0625] and near Cwm Celli [7945 1093, 7868 0832] provide well-exposed sections and south of the Pennal Fault it is well exposed on Pen yr Allt [7490 0110] and near Craig yr Ogof quarries [7376 0027]. However, the two best sections occur in the south-east and north-east corners of the outcrop, around Forge and Mynydd Hendre-ddu respectively.

In the Aberystwyth district (Cave and Hains, 1986, fig. 25), the formation is thickest, about 600 m, in the south and thins to 300–340 m in the north-east. In the Dyfi Forest area of the Cadair Idris district, the main area of outcrop, thickness estimates are inhibited by folding, but a section at the north end of the Dulas valley [7540 0630] indicates a thickness of 450–500 m. In the south-east of the district, the section east of Forge exposes 550 m of strata, although this includes an interdigitation, 170 m thick, of sandstone-poor strata comparable to the Borth Mudstones Formation (Figure 27). Towards the west, the sandstone component decreases and the formation thins markedly, to as little as 10 m around Happy Valley, and interdigitates with the Borth Mudstones Formation. This facies and thickness change broadly coincides with the Dyfi Syncline, the northward extension of the 'Glandyfi Tract' of the Aberystwyth district (Cave and Hains, 1986).

In the south and south-west of the district, the base of the formation is marked by turbiditic sandstones, some 0.05–0.15 m thick, and an increase in the thickness of mudstone turbidites from those of the underlying Cwmsymlog Formation. West of Pantperthog [7315 0434] and near Llugwy [7145 0013], the base is placed where the 20–30 mm thick sandstones of the Cwmsymlog Formation increase to 0.1 m. Elsewhere, the basal sandstones are poorly developed and the base is marked by a less pronounced increase in thickness of sandstones and turbiditic mudstones and the development of abundant phosphate layers.

The formation mainly comprises graded, fine-grained sandstone to mudstone turbidites, 0.08–0.3 m thick (Plate 11c). The sandstone is commonly ripple cross-

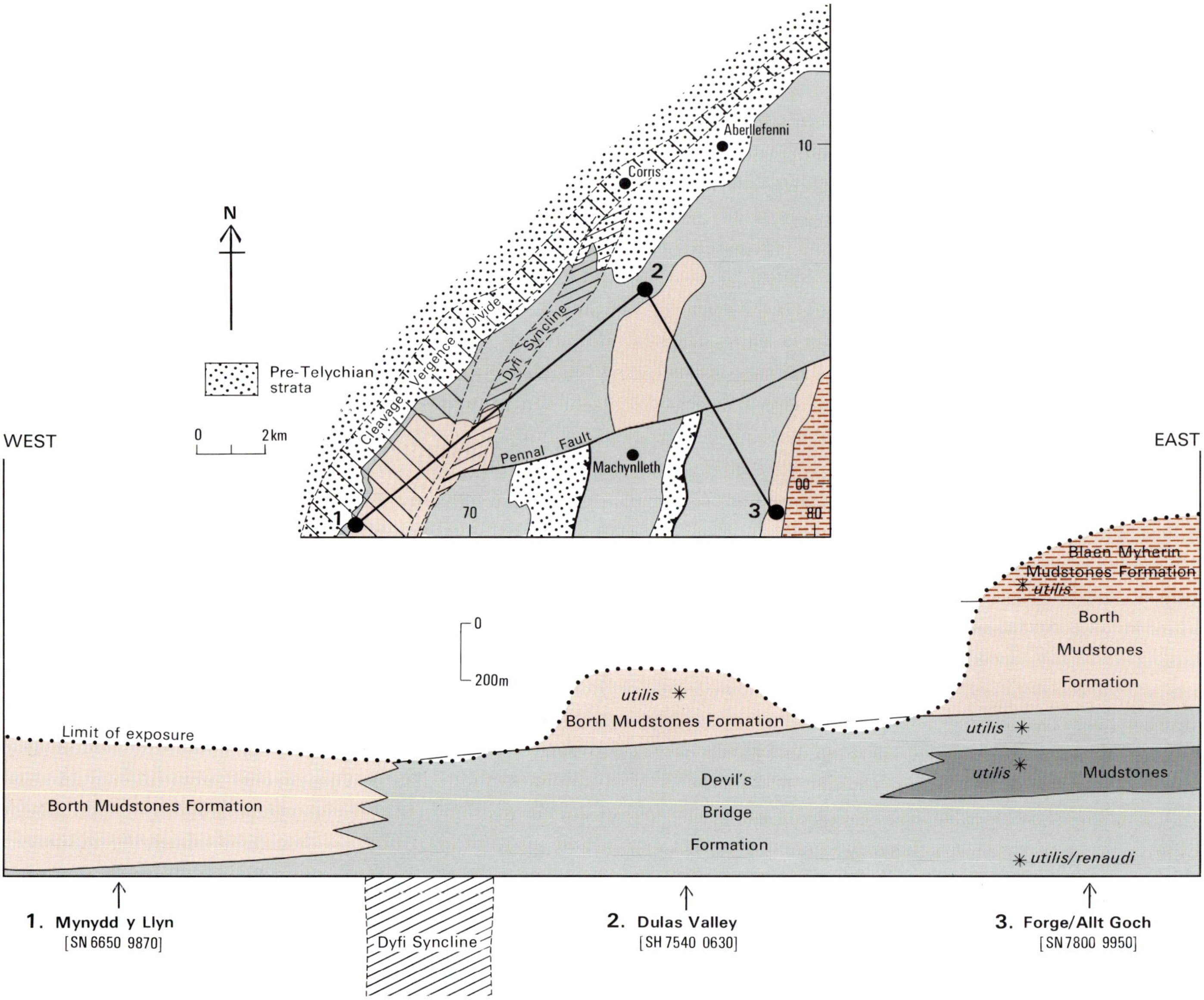

Figure 27 Fence diagram showing the relationships between the Telychian formations and the stratigraphic horizon of fossil collections. Map shows the outcrop of Telychian formations and the positions of the Dyfi Syncline and cleavage vergence divide.

stratified and grades up into parallel-laminated sandstone and a thicker silty mudstone portion with faint lamination. Many sandstone bedding planes, as those at Machynlleth railway station [7458 0135], are ripple-marked. Coarse-grained, graded sandstones are rare and generally contain a greater proportion of mudstone matrix, for example 0.1 m-thick beds on Cefn-Caer [7010 0341]. The trace fossils *Nereites*, *Dictyodora*, both grazing trails, and *Palaeodictyon*, a shallow burrow system, are locally preserved as casts on the bases of sandstone beds. The underlying hemipelagic mudstones commonly display *Chondrites*-type burrows infilled with sandstone. Bioturbation is particularly well displayed near Esgairlwyd [7804 0733], and east of Aberllefenni [8007 1050].

The sandstones are generally dominated by angular quartz grains, with subordinate feldspar, lithic fragments, muscovite and detrital chlorite-mica stacks (cf. Dimberline, 1986; Milodowski and Zalasiewicz, 1991a). The stacks are commonly concentrated, up to 40 per cent, in the ripple cross-stratified part of sandstone beds, where they alternate with quartz-rich bands.

The hemipelagic mudstones are mainly thin, 5–40 mm, and brown weathered with *Chondrites*-type burrows and locally with a faint lamination. Rare beds of laminated, dark grey and black hemipelagic mudstone, generally less than 10 mm thick, also occur.

Ripple cross-stratified sandstone, slump folds and rare flute marks indicate flow from the north-east and east and, less commonly, the south-east. At Cwm Cadian

[7364 0605], near the base of the formation, ripple cross-stratification indicates north-easterly derived palaeocurrents and recumbent slump folds suggest west-dipping palaeoslopes. Flute marks, at Blaen y glesyrch [7846 0803] and south of Nant Llidwy [7274 0465], indicate current flow from the north-east. In a quarry immediately south of the Afon Dyfi [7879 0238], palaeocurrent data from sandstones, up to 60 mm thick, also indicate derivation from the east and north-east.

Cone-in-cone nodules, up to 0.3 m diameter and 40–60 mm thick, of early diagenetic origin (Woodland, 1964; Craig, 1985), are common throughout the Devil's Bridge Formation. They are aligned parallel with bedding and are developed at the bases of sandstone beds or within the mudstone part of the turbidites. They comprise mainly quartz and iron-carbonate and, as near Pant Perthog [7465 0512], many are weathered out with an earthy, goethite deposit and elliptical impressions on the adjacent rock, for example south-east of Mynydd Rhyd-galed [7080 0385], north of Afon Rhonwydd [7166 0301, 7164 0289] and at Ty'n y Maes [7798 0070].

The formation is best exposed east of Forge, and around Penegoes, particularly in the Nant Dulas [SN 7650 9962], where it comprises alternating sequences, 5–20 m thick, of sandstone-poor and sandstone-rich turbidites, 0.05–0.2 m thick. These alternations form well-featured topography, as at Copa Shon [SN 7810 9930] and east of Penegoes [7895 0120]. At Pen y Graig-fawr [SN 7728 9952], approximately 250 m above the base of the formation, 170 m of sandstone-poor strata, similar to, and possibly an intercalation of, the Borth Mudstones Formation, is exposed in a broad synclinorium (Figure 27). These mudstone-dominated turbidites, darker grey than most Devil's Bridge Formation mudstones, are thicker, 0.3–0.5 m, and contain only very thin sandstones. Similar, and probably equivalent, beds are exposed at Machynlleth railway station [7458 0135], and comprise mudstone turbidites, 0.1–0.5 m thick, with basal sandstones, 40–50 mm thick.

Around Mynydd Hendre-ddu, in the north-east of the outcrop, the lowest 430 m of the formation is exposed in forestry tracks and roads. Between Llwydiarth [7916 1038] and Mynydd Hendre [8019 1179] the lowest strata are thinly bedded sandstones and siltstones, 5–40 mm thick, overlain by mudstone turbidites, 20–80 mm thick. The lowest 100 m of the formation contain at least two sequences of turbiditic mudstones and sandstones characterised by thin dark grey hemipelagic mudstones and siderite concretions, for example at [7922 1161]. To the north [8030 1120], minor scarps and hollows probably reflect the same sequence.

South of Mynydd Hendre-ddu [8020 1070], a road section exposes strata between 150 m, and 430 m, above the base of the formation. The lowest 20 m of the section comprise poorly cleaved mudstone turbidites, without sandstone bases, and thin bioturbated, hemipelagic mudstones; these beds are similar to the lithologies within the Borth Mudstones Formation. Rare sequences, up to 2 m thick, of thin, fissile, dark grey, hemipelagic mudstones are also exposed. Above, dark grey, hemipelagic mudstones are scarce and the thickness of individual mudstone turbidites progressively declines and sandstones thicken to 20 mm. The top 150 m of the succession, well exposed along a forestry track [8003 1036] to the south of the road, comprises turbidite rhythms, 20–50 mm thick, dominated by sandstones. Phosphate nodules and both vertical and horizontal bioturbation traces are widespread.

Conditions of deposition

The Devil's Bridge Formation was deposited within the relatively short time span of one or two graptolite subzones (*renaudi/utilis*) and marks an increase in sedimentation and subsidence rates. This change was probably initiated by the onset of major extensional faulting within, and along the east margin of the Welsh Basin (Wilson et al., 1992; Davies et al., in press), although Cave and Hains (1986) related it to temporary reversion of the Llandovery transgression, with a resultant increase in sediment supply.

The turbidites, the absence of shallow-water bedforms, and the trace fossil assemblages, which are similar to those of the Aberystwyth Grits (Crimes and Crossley, 1980; Cave and Hains, 1986), indicate accumulation below wavebase. Bioturbation and common phosphate concretions indicate mainly oxygenated bottom waters and oxidising conditions during early diagenesis. The rare laminated hemipelagic mudstones probably represent hiatuses in sedimentation and stratification of the water column, and may correlate with minor pulses during the overall Llandovery eustatic transgression (Ziegler et al., 1968; Bridges, 1975), when sediment supply was reduced by flooding of the shelf.

Palaeocurrent data indicate that the formation was derived mainly from the east and north-east, and is consistent with its thickest development in the east of the district (Figure 27). However, some turbidity currents also flowed from a south-easterly quadrant (Jones and Pugh, 1916; Cave and Hains, 1986).

Borth Mudstones Formation

The Borth Mudstones Formation, up to about 300 m thick, was formally defined by Cave (1975) and comprises turbiditic mudstones, in beds 0.2–0.6 m thick, with few sandstones. It crops out around Pennal, in a synclinal outlier in the Dulas valley [7658 0731–7681 0563] and south-east of the Dyfi valley (Figure 27). The best exposures are in quarries, such as those near Ceinws [7594 0538, 7671 0609 and 7603 0520], at Llwyn-gwern [7591 0448] and north-west of Rhos farch [6877 0117]. It is also well exposed, and is well featured, at Copa Shon, east of Forge [SN 7824 9928].

Within the Cadair Idris district, the formation interdigitates with the easterly derived Devil's Bridge Formation and is of similar age (*renaudi/utilis*) (Figure 27). The top of the formation is not exposed in the western part of the district, but in the east the formation is overlain by the Blaen Myherin Mudstones Formation. Because of their interdigitation, the boundary between the Devil's Bridge and Borth Mudstones formations is ill defined, and is placed where the mudstone turbidite beds are markedly thicker and the sandstones are thinner. It is best exposed north of Ceinws [7582 0637], where the mudstone turbidites increase upwards from 0.1–0.2 m to 0.4–0.5 m thick and east of Forge [SN 7824 9928].

The dominant lithology is fissile, medium-grey turbiditic mudstone, in beds 0.2–0.6 m thick. The mudstones are homogeneous or faintly laminated, with laminae defined by widely spaced, discontinuous siltstone streaks, only a few grains thick. Thin sandstone and siltstone beds, with ripple cross-stratification and planar lamination, occur at the base of some beds, but rarely amount to more than 10 per cent of the turbidite beds. Grazing trails on the bases of sandstones are less common than in the Devil's Bridge Formation. Hemipelagic mudstones, 20–30 mm thick, are mainly pale grey and bioturbated. Phosphate concretions occur, but are less common than in the Devil's Bridge Formation. Typical hemipelagic mudstones, approximately 100 m above the base of the formation, are well exposed in a quarry north of Ceinws [7594 0636].

In the Afon Dulas [7503 0375], less common, dark grey, laminated, hemipelagic mudstones, 1–20 mm thick, are well exposed. Nearby [7505 0383], a sequence, about 6 m thick, rich in dark grey mudstones, is exposed, in which mudstone turbidites are typically 0.2–0.6 m thick and are overlain by 30–40 mm of laminated, graptolitic dark grey mudstone. Ellipsoidal monazite nodules (Milodowski and Zalasiewicz, 1991b), up to 1 mm in length, and pyrite nodules are scattered within the hemipelagic mudstone, and, less commonly, within the turbiditic mudstone.

Conditions of deposition

The formation represents sedimentation from low-density, distal turbidity currents, dominantly in oxygenated bottom waters. Although there are no palaeocurrent data, a large part of the formation in the west of the district is equivalent to the easterly derived Devil's Bridge Formation (Figure 27). However, there was probably additional input from the south since the formation is also contemporaneous with the southerly derived Aberystwyth Grits Group (Cave and Hains, 1986; Loydell, 1991).

The interdigitation of the Devil's Bridge and Borth Mudstones formations across the Dyfi Syncline suggests a tectonically induced control on sedimentation and bathymetry. In the Aberystwyth and Llanilar districts, Wilson et al. (1992) proposed that the possible southward continuation of the Dyfi Syncline and cleavage vergence divide was a syndepositional fault, with a downthrow to the west. A similar model may be applicable to the Cadair Idris district, but the simplest interpretation is that a syndepositional fault, with an easterly downthrow, constrained sandier, more powerful turbidity currents, represented by the Devil's Bridge Formation, to its eastern side. In the second model, only lower density turbidity currents, depositing the thicker, muddier beds of the Borth Mudstones Formation, crossed the east-facing step in the sea floor. The termination of the Dyfi Syncline against the outcrop of the Nod Glas Formation at Corris (Figure 28), suggests that the fault may have been listric, possibly rooted into the Nod Glas Formation at a depth of only 1–2 km below the sea bed.

Blaen Myherin Mudstones Formation

The Blaen Myherin Mudstones Formation (Jones, 1909; Cave and Hains, 1986) overlies the Borth Mudstones Formation and crops out only in the south-east corner of the district (Figure 27). It was previously referred to as the Crewi Formation (Cave and Scott, 1990). The formation, best exposed in the bed of the Crewi river [7911 0027], is differentiated from the Borth Mudstones and Devil's Bridge formations, which are mainly 'oxic', by darker, better-cleaved, turbiditic mudstones and an increase in the proportion of dark grey, laminated hemipelagic mudstones. A *utilis* subzone fauna has been located from near the base of the formation (Figure 27).

In a track north of Allt Goch [SN 7941 9986], the basal 55 m, including the contact with the Borth Mudstones Formation, is well exposed in a syncline. The strata comprise thickly bedded, turbiditic mudstones with black, laminated hemipelagic mudstones, up to 30 mm thick, spaced at 0.5 m intervals. Rare sandstones, up to 50 mm thick, are commonly accentuated by the development of cone-in-cone concretions from their bedding planes into the adjacent mudstones. Ripple cross-stratification indicates palaeocurrents from the north-east.

The top of the formation does not crop out within the district and the maximum thickness, exposed on Ffridd Wyllt [7980 0079], is about 300 m, where the highest beds include a greater proportion of sandstone and rare bioturbated hemipelagic mudstone with phosphate concretions. South of Abercegir [8001 0100], two bioturbated, muddy, medium-grained sandstone beds, each 0.7 m thick, with mudstone rip-up clasts up to 0.2 m long, form distinctive dip slopes.

Conditions of deposition

The formation was deposited under fluctuating, but mainly poorly oxygenated, bottom waters and marks an end of the typical, oxygenated, Devil's Bridge/Borth Mudstones regime. The thick sandstones at Ffridd Wyllt may be related to the southerly derived Cwmystwyth Grits Group of the Rhayader/Llanilar districts (Davies et al., in press).

FIVE

Intrusions

Intrusions are widespread within the Cambrian and Ordovician strata, particularly within the basic volcanic formations, but none occurs in strata above the Aran Volcanic Group. Geochemical evidence and field relationships, including evidence of intrusion into unlithified sediments, suggests that most intrusions were contemporaneous with volcanism (Allen et al., 1976; Dunkley, 1978). In the Harlech district two phases of intrusion, related to the Rhobell (Tremadoc) and Aran Volcanic groups (Arenig–Caradoc), have long been recognised (Ramsay, 1881; Wells, 1925; Allen et al., 1976; Kokelaar, 1977, 1979). However, within this district, intrusions related to the Rhobell Volcanic Group are less common.

Dolerite

Typically the dolerites are massive, well jointed and green-grey or rusty weathered, but thin intrusions are locally weakly cleaved. The majority of intrusions are sills. Columnar jointing is widespread in the central portions of sills and a few display flow banding. The lower parts of the sills are generally coarser grained and the intrusions at the lower stratigraphic levels are commonly thicker, coarser grained and more mafic, than those at higher levels. Chilled margins in even the largest intrusions are rarely more than a few centimetres thick and the surrounding hornfelsed zone is rarely more than a few metres thick.

In the west of the district, dolerite intrusions are restricted to strata below the Llyn y Gafr Volcanic Formation. However, to the east, intrusions occur at progressively higher levels, with the highest occurring in a mudstone interbed of the Craig Cau Formation [7597 1304], within 100 m of the top of the Aran Volcanic Group.

The thickest dolerite sills, up to 350 m thick, intrude the Allt Lŵyd, Offrwm Volcanic and Cregennen formations in the vicinity of Dolgellau and include the 'Mynydd y Gader Dolerite' of Lake and Reynolds (1896, 1912) which crops out between Tyddyn-mawr and Afon Aran [7033 1535 to 7392 1560]. Similar, large, locally gabbroic, intrusions can be traced from Tonfanau to Llwyngwril and farther north-eastwards to Arthog. Dolerite sills are also common within the Llyn y Gafr and Pen y gadair Volcanic formations. The 'Pen-y-gader dolerite' of Davies (1956), about 90 m thick, is well exposed immediately west of the summit of Penygadair [7122 1320] and can be traced eastward to the Mynydd Moel microgranite boss. An equivalent dolerite on the east side of the boss, locally up to 220 m thick, can be traced as far as Cross Foxes [7665 1669]. As with many of the intrusions, the Pen-y-gader dolerite was emplaced into mudstone, suggesting that mudstone layers facilitated intrusion.

Near Llyn Wylfa [6709 1598], small mudstone 'flames', less than 20 mm across, intrude the lower, chilled margin of a thin (3.75 m) dolerite sill, indicating that the sediment was unlithified at the time of intrusion. Near Gefnir Farm [6580 1515], larger mudstone 'flames', up to 1 m, penetrate a dolerite sill.

At Tonfanau Quarry [5720 0340] a dolerite sill intrudes the Offrwm Volcanic Formation. The intrusion is broadly concordant, but its margins are locally irregular, with thin apophyses and dykes in the adjacent tuffs. The dolerite contains numerous small xenoliths of hornfelsed mudstone and rafts of tuff, up to 70 m long and 10–15 m thick, aligned parallel to the margins of the intrusion [5721 0346]. Tuff rafts, up to 30 m in length and 3 m thick, also occur south of the quarry [5722 0311]. In the highest gallery of the quarry, and to the east at Tal-y-Gareg [5743 0353], coarse-grained dolerite is intruded by a stockwork of fine-grained dolerite, with narrow (about 10 mm) chilled margins, indicating multiple intrusion (Plate 12a).

Petrography

The dolerites are predominantly medium grained and non-porphyritic with subophitic and/or equigranular textures. Most comprise a groundmass of interlocking, zoned albite/oligoclase crystals, probably albitised calcic plagioclase, and interstitial chlorite with ophitic plates, up to 10 mm across, of pale pink, pleochroic, titaniferous augite (E62610, E67072) (Plate 14a). Some pyroxenes contain rounded inclusions of chlorite and actinolite, probably after olivine (E67072). Skeletal crystals of sphene and leucoxene with interlocking exsolution lamellae, probably altered ilmenite, are widespread (E65175). Rare simple twinned potassium feldspars, forming elongate prisms, occur in the margin of the Tonfanau intrusion. Quartz is a minor component of many dolerites, particularly in the margins of intrusions. Its interstitial occurrence, and tendency to form euhedral crystals within vugs, indicates late-stage crystallisation.

The dolerites are variably altered, with the central portions of sills generally less affected than the margins. Locally, intense alteration, with complete replacement of pyroxene by actinolite and chlorite, has obscured the primary texture (E62097, E61105). However, the primary igneous texture is discernible in most examples, although the peripheries of augite crystals invariably show some secondary actinolite and chlorite and feldpars are commonly sericitised and epidotised (E65031, E62088, E61613). Chilled margins are locally vesicular with feldspar phenocrysts in a groundmass of feldspar microlites and chlorite (E65129).

Plate 12a Coarse-grained dolerite intruded by finer grained dolerite with chilled margins (A14579). Tonfanau 'granite' quarry [5720 0340]. Field of view about 0.75 m.

Plate 12b Foliated dolerite intrusion. North of Mynydd Moel [7273 1423].

Basalt

Concordant sheets and pods of basalt, up to 150 m thick, are widespread within the Llyn y Gafr Volcanic, Pen y gadair Volcanic and Benglog formations. They are texturally and mineralogically similar to the intrusive dolerites, although finer grained, and indistinguishable petrographically from the coarse-grained portions of lava flows. Without well-exposed contacts it is difficult to determine whether the sheets are intrusive or extrusive. Some intrusions probably pass laterally into extrusions.

South of Arthog [6369 1279], pod-like bodies of basalt within Cregennen Formation mudstones are difficult to interpret. Contacts with the mudstones are not exposed, but the pods, up to about 20 m thick, are possibly high-level intrusions into unlithified sediment. Locally, the basalts are columnar jointed and flow banded.

Between Tyrrau Mawr and Mynydd Moel, basalt sheets, up to 100 m thick, are also difficult to separate from the adjacent massive lavas of the Llyn y Gafr Volcanic Formation. North of Mynydd Moel, the basalt sheets are columnar jointed [7255 1455] and locally foliated [7239 1439] (Plate 12b). Some, with upper, chilled margins, 0.05–0.15 m thick, are demonstrably intrusive [7256 1454]. At Bwlch Oerddrws [7940 1705 to 7972 1700], mudstones interbedded with crystal tuffs of the Benglog Formation facilitated the emplacement of basalt sheets, 2–20 m thick. The irregular form of some sheets [7944 1704] and mudstone flames along the upper contact [7957 1698] suggest intrusion into unlithified sediments.

South of Dolgellau [7193 1611 to 7298 1570], porphyritic basalt, up to 40 m thick, within the Dolgellau Formation, was referred to as 'andesite' by Lake and Reynolds (1912) and Cox and Wells (1921). It is highly altered and comprises an aggregate of calcite, chlorite and granular feldspar with highly altered phenocrysts, possibly after plagioclase (E65935, E65938).

The only other porphyritic basalts occur within the Offrwm Volcanic Formation near Arthog [6574 1520], and at the top of the dolerite at Mynydd-y-Gader [7292 1499]. The latter is intruded by pale weathered basalt sheets up to 0.3 m thick. Typically, the basalts comprise phenocrysts of altered plagioclase (10–20 per cent) in a groundmass of felted feldspar laths (E65921, E62033). Calcite and chlorite pseudomorphs of mafic phenocrysts, and iron oxide microphenocrysts, are also common (E65956).

Microdiorite

Microdiorite intrusions, mainly sills, are related to the activity of the Rhobell Volcanic Group (Tremadoc) and mostly occur in the north-east of the district. Most intrusions occur within the Dolgellau Formation, particularly in the outcrop between Arthog and Dolgellau, suggesting that the mudstones allowed easy access for the magma. South-east of Fairbourne, at the 'Blue Lake' quarry [6215 1210], two sills, 0.06–0.3 m and 5 m thick, intrude the Dolgellau Formation. The thicker sill is columnar jointed, and at its lower contact there is about 1.5 m of spotted, hornfelsed mudstone. Between Arthog and Dolgedr [6965 1740] thicker microdiorite sills, up to 300 m thick, occur within the Ffestiniog Flags, Dolgellau and Dol-cyn-afon formations, and are locally discordant [6535 1593, 6891 1746]. South of Dolgellau, microdiorite sills, 10–20 m thick, occur within the Ffestiniog Flags Formation [7344 1702, 7400 1736]. In the coastal section north of Llwyngwril [5994 1107, 6008 1125], a microdiorite sill, about 20 m thick, and a dyke, 2.5 m thick, intrude the Gamlan Formation (Figure 32).

Petrography

The microdiorites are mainly aphyric and equigranular. Albitised plagioclase is abundant and commonly altered to sericite, chlorite and calcite (E62025, E61106, E65194). Original mafic minerals are completely pseudomorphed by aggregates of chlorite, calcite and iron oxide (E62025). Interstitial quartz is common (E62035, E62027, E65585) and in places granophyric intergrowths of quartz and potassium feldspar (E62078) can be determined in the groundmass.

Microgranite and rhyolite

Acidic intrusions are common within the Aran Volcanic Group around the Harlech Dome (Lynas, 1973; Dunkley, 1978, 1979; Allen and Jackson, 1985). Within the district, the Cregennen and Cadair Idris microgranite intrusions are the largest and best exposed (Davies, 1955, 1959; Phillips, 1966). The intrusions, up to 600 m thick, are mainly concordant but with boss-like, cross-cutting terminations and are interpreted as high-level intrusions emplaced during the late stages of the Aran Volcanic Group volcanism. The microgranites are typically white or pink weathered, compact, massive or columnar jointed (Plate 13a) and generally form rugged terrain. In thin section, scattered albite-oligoclase phenocrysts, up to 4 mm across (E65778), within a fine- to medium-grained, granophyric groundmass can be distinguished (Plate 14b). Biotite, chlorite, zircon, sphene and apatite are the main accessories (E62071, E65779).

The Cregennen sill, up to 500 m thick, is slightly transgressive and between Llynnau Cregennen and Gellilwyd [7071 1672] it intrudes the Dol-cyn-afon, Offrwm Volcanic, Cregennen and Llyn y Gafr Volcanic formations. Its western termination is obscured by the drift cover, although a small pod at Braich Ddu [6426 1210], about 15 m thick, and a larger, faulted mass of microgranite south of the Ceunant Fault [6700 1350], may be the lateral equivalents. The microgranite is mainly massive, poorly jointed and homogeneous, but is locally flow banded [6610 1466] and brecciated [6710 1553].

The Cregennen microgranite is best exposed on Pared y Cefn-hir [6610 1469 to 6670 1520] (Plate 4c), where the lower margin is distinct and basic in composition (Phillips, 1966). At the contact, hornfelsed mudstones are overlain by a dark green-weathered, fine-grained quartz dolerite, 5 m thick, with granophyric clots, a few sodic plagioclase phenocrysts, actinolite, biotite/stilpnomelane, opaque oxide, apatite, sphene and zircon (E62066–7). The dolerite grades up [6651 1506] into a concordant layer, 3 m thick, of microgranite with dolerite xenoliths, up to 0.2 m across, which decrease in size and abundance upwards and are locally contorted (Plate 13b). The contacts of the xenoliths with the microgranite are diffuse with less amphibole and biotite. The basic margin is persistent along of the lower contact of the Cregennen sill and is also well exposed at Pen Moelyn [6686 1544 to 6703 1560], Gallestra [6785 1596], Kings [6853 1619] and Tal-y-waen [6934 1691].

The basic margin of the Cregennen microgranite is not considered to represent in-situ fractionation, particularly since other acid intrusions within the district display similar, basic upper margins. Furthermore, doleritic xenoliths occur throughout the Braich Ddu pod, which contains no basic margin, and therefore must have been transported from their original site of crystallisation. Instead, the xenolithic zone of the Cregennen microgranite suggests disruption of an earlier, semi consolidated basaltic magma, emplaced before the main mass of the microgranite.

The Cadair Idris sill, up to 600 m thick, crops out over 4.75 km between Craig-las and Mynydd Moel and mainly intrudes strata above the Fron Newydd Member. In the west, near Craig-las [6818 1385], the sill wedges out within mudstones, but to the east, at Mynydd Moel [7275 1365], it is rhyolitic and markedly discordant within the Pen y

Plate 13a Columnar jointing in the Cadair Idris microgranite. North side of Mynydd Moel [7274 1367].

Plate 13b Basic inclusions in the margin of the Cregennen microgranite [6651 1506].

gadair Volcanic, and Craig Cau formations. For much of its outcrop, the lower contact is obscured by scree, but on the north and north-west sides of Mynydd Moel [7199 1388 to 7268 1402], a basic marginal zone, up to 15 m thick, overlies hornfelsed crystal tuff. On the north, north-east and south-east sides of Mynydd Moel the lower part of the microgranite is banded [7318 1344], with brown-grey to pink-grey, fine-grained bands, 10–20 mm thick, and coarser, paler grey bands 3–20 mm thick.

The Cadair Idris microgranite is characterised by spectacular columnar joints, as at Cyfrwy [7040 1345], Pen-y-gadair [7135 1335], Twr Du [7190 1350] and Mynydd Moel [7280 1370] (Plate 13a). Its upper contact is prominently featured in the cliffs around the summit [7125 1327], and farther east it is exposed near Twr Du and Mynydd Moel [7204 1344]. A basic upper margin is locally [7204 1344] distinguished, but elsewhere the chilled margin [7207 1345] contains siliceous nodules up to 30 mm across. In places [7131 1329], contorted bedding in the adjacent hornfelsed mudstones is attributed to the effects of the intrusion.

The eastern edge of the Mynydd Moel boss is clearly defined between [7314 1355] and [7335 1306], where it transects the Pen y gadair Volcanic and Ty'r Gawen Mudstone formations, and the lower part of the Craig Cau Formation. However, at Craig Cwmrhwyddfor, the microgranite grades imperceptibly into rhyolite which intrudes the uppermost ash-flow tuff of the Craig Cau Formation

Plate 14a Plate of clinopyroxene in coarse-grained dolerite displaying subophitic relationships with feldspar crystals (E67072). Tonfanau [5764 0386]. Field of view is about 1.5 mm wide.

Plate 14b Granophyric texture in microgranite. Cadair Idris microgranite [7045 1318]. Field of view is about 6 mm wide.

[7350 1300 to 7385 1290] and the contact is difficult to distinguish because of the similarity of the recrystallised ash-flow tuff and the rhyolite. Flow banding is locally developed [7364 1297] and is defined by alternating chlorite-rich and chlorite-poor bands (E65788) and granophyric patches are common (E65786). It is possible that the Mynydd Moel rhyolite breaches the Craig Cau Formation, as maintained by Davies (1959), since at Nant Caenewydd [7342 1204] possibly extrusive rhyolite is locally in contact with unbaked mudstones of the Ceiswyn Formation.

The upward transition of microgranite to rhyolite also occurs on Mynydd Pennant [6600 1100], where a broadly concordant microgranite intrusion trangresses into a rhyolite within the Craig Cau Formation. However, poor exposure inhibits clear distinction of the form of the rhyolite and its relationship with the microgranite sill. The pink-weathered, feldsparphyric, flow–banded rhyolite is best exposed in stream sections [6663 1087] some distance away from the microgranite. In thin section, spherulitic recrystallisation and albite and potassium feldspar phenocrysts, up to 2 mm in length, are distinguished (E61634). Closer to the microgranite [6619 1100], the foliated rhyolite is highly recrystallised, but contains ghost granophyric textures (E66367).

Concordant sheets of rhyolite, probably intrusions, also occur within the central part of the outcrop of the Craig Cau Formation, between Craig Cwmrhyddfor and Craig y Llam, although they have been noted as far west as Castell-y-bere. However, their field relationships are difficult to determine because they are so lithologically similar to the acid tuffs and in many instances the original fabric has been obliterated by recrystallisation.

Small pods of rhyolite/dacite intrude the Cregennen and Llyn y Gafr Volcanic formations at Llanegryn [6063 0685] and within the Gwril valley [6204 0892]. Their contacts with the country rock are not exposed, but evidence of contact alteration suggests they are intrusive. The Llanegryn intrusion is almost circular in outcrop, with a diameter of 40 m, and comprises pale-weathered, massive rhyolite with albite phenocrysts, up to 3 mm long, and blebs of quartz and chlorite (E67027). The Gwril intrusion, exposed only in the stream bed, is sill-like and comprises dark green weathered rhyolite with flow-aligned albite phenocrysts up to 5 mm long. In thin section, flow-foliated feldspar microlites can be distinguished and the mafic content is very low (E65808).

Contact metamorphism and metasomatism

The contact altered zones, adjacent to even the thickest dolerite intrusions, rarely exceed 2 m in width and are commonly only a few centimetres wide. Typical is a 3 m wide aureole around a dolerite intrusion, 120 m thick, near Gefnir Farm [6582 1512] which comprises splintery, pale grey hornfelsed mudstone. However, within the lower part of the Aran Volcanic Group, the dolerites are generally coarser and more mafic, and wider aureoles have been distinguished. The zone around the Tonfanau dolerite is up to 20 m wide and at Llanegryn [6042 0637] 15–20 m of mudstones between two dolerite sills are hornfelsed and spotted throughout (E67029). Contact altered mudstones, such as those at Tonfanau Beacon [5743 0353], are generally pale grey or green with deformed spots, up to 0.5 mm in diameter, of recrystallised quartz which are relatively depleted in chlorite (E67021, E67070, E67090). The long axes of the spots are coincident with a weak S_1 alignment of white mica flakes. Veinlets of quartz with euhedral sphene also occur. Spotted mudstones from Bodowen [5860 0540] contain prismatic porphyroblasts pseudomorphed by stilpnomelane and chlorite, possibly after andalusite.

Tuffs adjacent to the Tonfanau dolerite intrusion [5767 0391] contain hydrothermal veins of albite and quartz with euhedral clinozoisite and sprays of stilpnomelane (E67073). Within the tuffs, plagioclase crystals are highly altered, both sericitised and, 1 m away from the contact [5717 0317], largely replaced by calcite and pyrite. Geochemical analyses of hornfelsed mudstones around the 'Pen-y-gader dolerite' on Cadair Idris also imply some Na, Bo and F metasomatism of the country rocks (Davies, 1959).

SIX

Structure

The Lower Palaeozoic rocks of the Welsh Basin were deformed and metamorphosed during the end-Caledonian (Lower Devonian) orogeny (Jones, 1956) at the 'hard' collision between Laurentia and Avalonia and the closure of the Iapetus Ocean (Soper et al., 1987). Within the district, upright folding, cleavage development and reverse faulting shortened the Lower Palaeozoic sequence by 50–60 per cent. Fold development was mainly controlled by contrasts in competence between various stratigraphical elements. However, faulting rooted within ductile layers such as the Nod Glas Formation also exerted a strong influence.

The affects of the Variscan (late Carboniferous) deformation, caused by the closure of the 'Proto-Tethys' ocean to the south, were mild and were probably restricted to fault displacements, some of them strike-slip. The net normal dip-slip displacements on many of the faults within the district are probably younger, perhaps associated with the extension and subsidence of the Mesozoic basin in Cardigan Bay.

Folds

The Caerdeon Syncline, the only fold that can be related to the major folds of the Harlech district, becomes disharmonic southwards, and it can only be distinguished in the outcrop of the Aran Volcanic Group for about 3–4 km south of the Mawddach estuary (Figure 28). The other major fold in the north of the district is the open Dol-Ithel Anticline, which affects strata ranging from the Ffestiniog Flags to the Ceiswyn formations. It can be traced from Tal-y-llyn to Dolgellau, and to the north it passes into the Rhobell Fracture Zone of the Harlech district (Kokelaar, 1988). It is likely that the fold developed on a pre-Caledonian structure because Tremadoc strata were removed by erosion in the hinge zone before deposition of the Aran Volcanic Group, suggesting that it was an axis of uplift in late Tremadoc–early Arenig times (Cox and Wells, 1921).

Minor folds, 10–100 m in wavelength, are common in the well-exposed Cambrian strata in the coast section between Llwyngwril [6010 1120] and Fairbourne [6105 1187] (Figure 6). Most of the folds are upright and open, with interlimb angles of more than 100°, and verge weakly to the north-west. Marked thickening in the hinge zones, largely accommodated in the mudstone beds, has resulted in similar style folds (Classes 1C and 2, Ramsay, 1967). Folds of competent beds, in particular the thickly bedded sandstones in the Clogau Formation, are tighter than those in more thinly bedded strata and steep limbs are commonly faulted out (Plate 15a).

Bedding is poorly defined in the Aran Volcanic Group mudstones and, as a result, folds are difficult to distinguish. However, folds can be determined between the Bird Rock Anticline and the Llanegryn Fault, in a tract of subvertical strata, about 1 km^2 in area, around Ty'r Gawen [6200 0650]. There, the mean fold axis, which plunges 60° to the east-north-east, is anomalous compared with the regional fold pattern (Figure 29). Rare mesoscale folds exposed at Ty'r Gawen [6179 0595] strike north-north-east and plunge nearly vertically.

The Craig Cau Formation is folded into a series of north–south-striking, south-plunging folds with a mean wavelength of 2 km (Figures 28 and 30). The folds maintain an almost parallel form (Class 1B, Ramsay, 1967), but a locally developed, weak S_1 cleavage within the tuffs indicates a small amount of flattening in the hinge zones. The fold- train indicates a minimum of 50 per cent shortening (Pratt, 1990). The form of the folds is largely determined by the thickness of the tuffs and the thickness of the underlying mudstones. Consequently, the Craig Ysgiog Anticline, in the east of the outcrop, has low amplitude, interlimb angles of 60–70° and, to the north, can be determined down to below the Pen y gadair Volcanic Formation. In contrast, the isoclinal, high amplitude Bird Rock Anticline, in the west of the outcrop, cannot be traced far above or below the Craig Cau Formation and appears to be 'rootless'. The thick sequence of surrounding mudstones probably accommodated the folding by ductile flow rather than *décollement*. Where the tuffs wedge out, major folds are not developed, suggesting that end-Caledonian folding nucleated on these competent beds.

The north–south-striking folds in the Craig Cau Formation cannot be traced far to the south and in the higher, mudstone-dominated strata they are succeeded by numerous shorter wavelength, open folds. Most folds plunge southwards, at 10–25°, but axial plane traces are deflected to a north-north-east–south-south-west orientation (Figure 28). Well-exposed folds occur around Foel Caerberllan [6750 0840], Mynydd Tyn-y-fach [6880 0950], Abergynolwyn [6705 0725] and Cwm Amarch [7060 1120]. The longest wavelength folds in the district, the Dol-Ithel and Mynydd y Llyn anticlines, which occur north and south of the Tal-y-llyn Fault respectively, are important fold vergence divides. To the west of these structures minor folds are generally M-shaped or south-east-verging, and to the east they are north-west-verging.

The Dyfi Syncline, a broad synclinorium, is also a major fold vergence divide and can be traced from Corris to the southern edge of the district, near Pennal; to the east folds verge south-east and are associated with eastwards directed, reverse faulting and thrusting, for example the Forge and Gelli-goch 'Overthrusts' (Figure 28). Folds in the south-east of the district are markedly periclinal and oriented closer to north–south. A broad depression in fold plunge, between the southerly plunging folds of the Harlech Dome and the northerly plung-

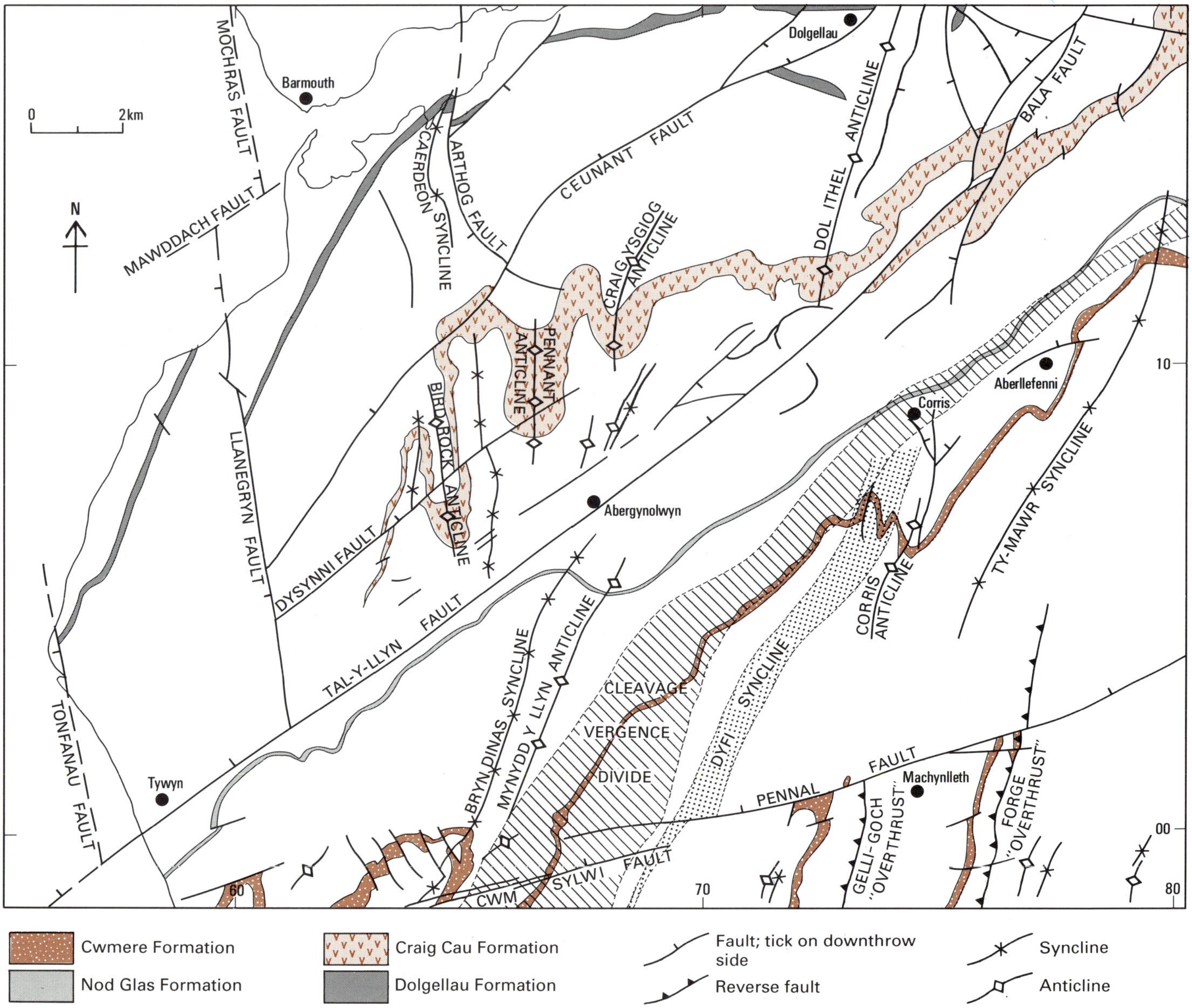

Figure 28 Main structures of the Cadair Idris district. The S_1 cleavage vergence divide is a tractof approximately vertical cleavage. The Dyfi Syncline is an important fold vergence divide.

ing folds of the Plynlimon Dome–Machynlleth Inlier, is poorly defined because of the complex interdigitation of periclinal folds but approximates to the Afon Dyfi.

Cleavages

The volcanic and intrusive rocks are weakly cleaved or uncleaved, but most of the sedimentary rocks display a strong cleavage (S_1). Its spacing, planarity and orientation was mainly determined by grain size, and less influenced by bedding. Well-bedded mudstones and siltstones are characterised by a spaced, planar cleavage, whereas disturbed beds, such as those of the Ty'r Gawen Mudstone, Ceiswyn and Garnedd-wen formations, have anastomosing, widely spaced cleavages.

In the black mudstones, such as those of the Clogau, Nod Glas and Cwmere formations, the S_1 cleavage approaches a 'continuous cleavage' (Powell, 1979) with a steeply plunging stretching lineation. Detrital grains and pyrite accumulations have associated pressure shadows of quartz and chlorite which indicate up-cleavage extension of at least 110 per cent (Pratt, 1990). In contrast, in the sandstone beds the S_1 cleavage planes are widely spaced and anastomosing. Sandstones with little matrix, and those with diagenetic quartz cements, are less well cleaved. The S_1 cleavage is strongly refracted at sand-

Plate 15a Strongly asymmetrical, tight fold within the Clogau Formation. Note the excision of the thick sandstone to the left of the fold by broadly bedding-parallel faults. Llwyngwril [6010 1123].

Plate 15b Kink bands within the Ceiswyn Formation beside the Tal-y-llyn Fault. Abergynolwyn Station [6698 0648].

stone–mudstone contacts and within graded sandstone–mudstone beds the traces are curved. Consequently, cleavage commonly fans about fold hinges in sandstone-rich lithologies. However, in mudstone-dominated formations the folds display either non-fanning cleavage, or weakly divergent fans.

In most of the district, the S_1 cleavage strikes between north-east–south-west and north-north-east–south-south-west and dips at more than 75° (Figure 29). However, the cleavage orientation is more variable close to the competent rocks and it is almost concordant with bedding in thin mudstones within the Aran Volcanic Group. Cleavage also fans around hinges of the folds of competent tuffs, such as the Bird Rock and Dol-Ithel anticlines.

A broad cleavage vergence divide between Happy Valley, in the south, and Corris, in the north, separates south-east-dipping cleavage to the west, from north-west-dipping cleavage to the east (Figures 28 and 29). This zone lies mainly 1–2 km west of the Dyfi Syncline and is probably the northern continuation of the 'Glandyfi Tract' of the Aberystwyth district (Cave and Hains, 1986). However, at Corris the zone narrows to about 1 km and coincides with the Dyfi Syncline. North-east of Corris, the cleavage vergence divide is coincident with the outcrop of the Nod Glas and Broad Vein formations.

A locally developed S_2 crenulation cleavage has been recognised in the vicinity of faults and tight folds, and

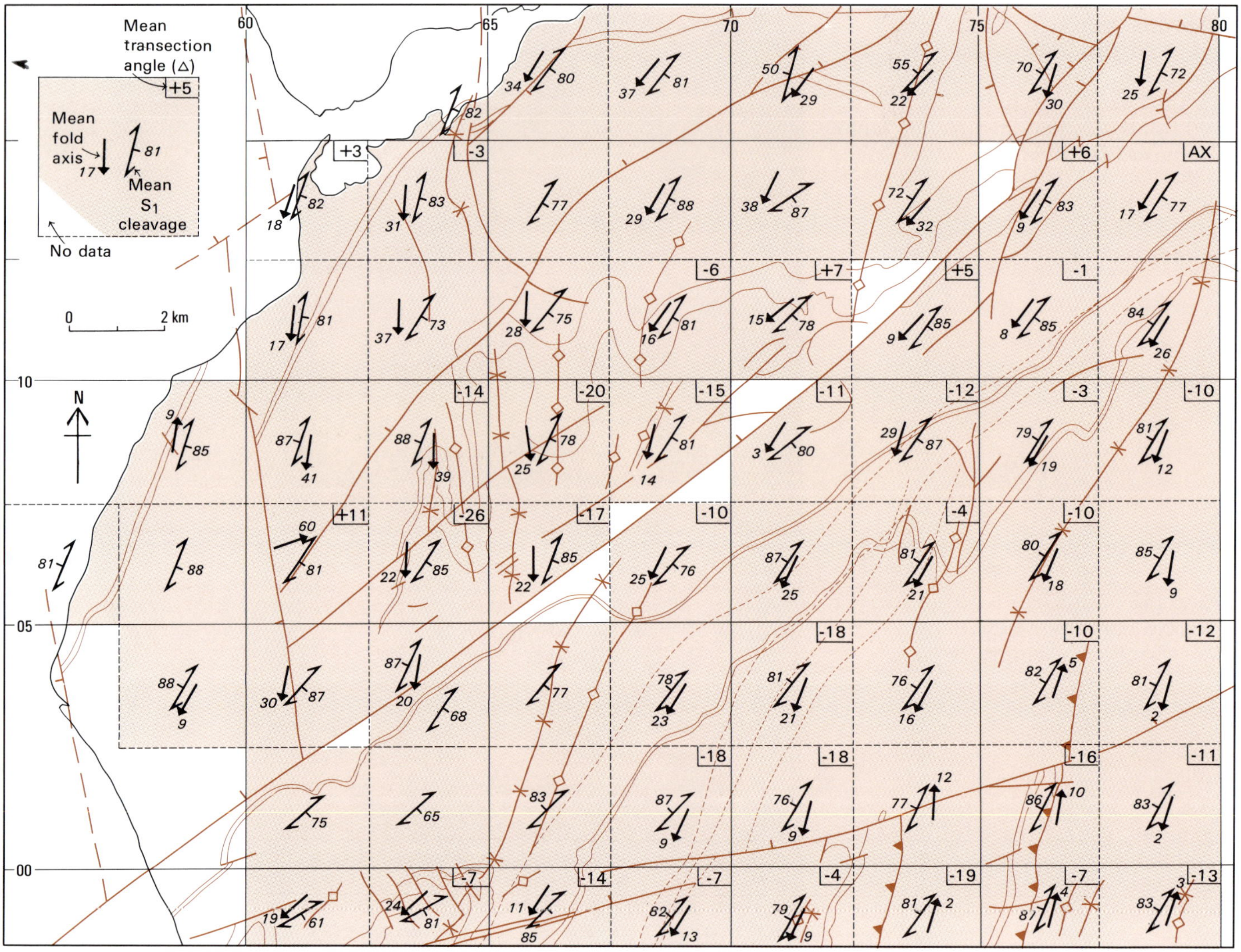

Figure 29 Synthesis of structural data (S_1 cleavage, fold axes) and transection angles (Δ of Borradaile, 1978) for 2.5 × 2.5 km subareas. Total of 8077 bedding and 6341 S_1 readings. Negative Δ values indicate clockwise transection, positive Δ values indicate anticlockwise transection (convention of Johnson, 1991). AX indicates axial planar cleavage. Transection angles are not given for those subareas with markedly periclinal folds or with limited data.

along the contacts between layers of markedly contrasting competence. In mudstones in the cores of synclines in the Craig Cau Formation, it crenulates a vertical S_1 cleavage, strikes almost east–west and generally dips south at 20–40°, approximately parallel to the plunge of the folds (Figures 28 and 30). The crenulations, spaced at about 0.5 mm, are asymmetrical and commonly terminate against chlorite-mica stacks, monazite nodules or detrital quartz grains (E67098). This S_2 cleavage may have developed during post-Caledonian, orogenic relaxation, when the maximum compressive stress was vertical, or possibly by dip-slip faulting along the nearly vertical contact between competent volcanic rocks and the incompetent mudstones.

Strike-slip movements along the Tal-y-llyn Fault zone also caused S_2 cleavage in drag folds of the S_1 cleavage. Within a few metres of a fault at Mynydd Pentre [7235 1122], steeply plunging drag folds of S_1 have an axial planar, steep to vertical S_2 cleavage, and are associated with nearly vertical kink bands. Similar cleavages and drag folds occur adjacent to east–west-striking faults on Foel Wyllt [6350 0531 and 6375 0506].

Around Ty'r Gawen, west of the Bird Rock Anticline, a vertical S_2 cleavage, spaced at 2–10 mm, strikes north–south (Figure 28). Its intersection with north-east-striking S_1 cleavage planes forms steeply plunging 'pencils' of mudstone. Pressure solution seams of iron oxides are pronounced within the hinge zones of the asymmetrical crenulations. The cause of this S_2 cleavage is uncertain; it is possible that it developed in response to strike-slip movement on the nearby Ceunant Fault. The S_2 cleavage also coincides with the anomalous folds described above,

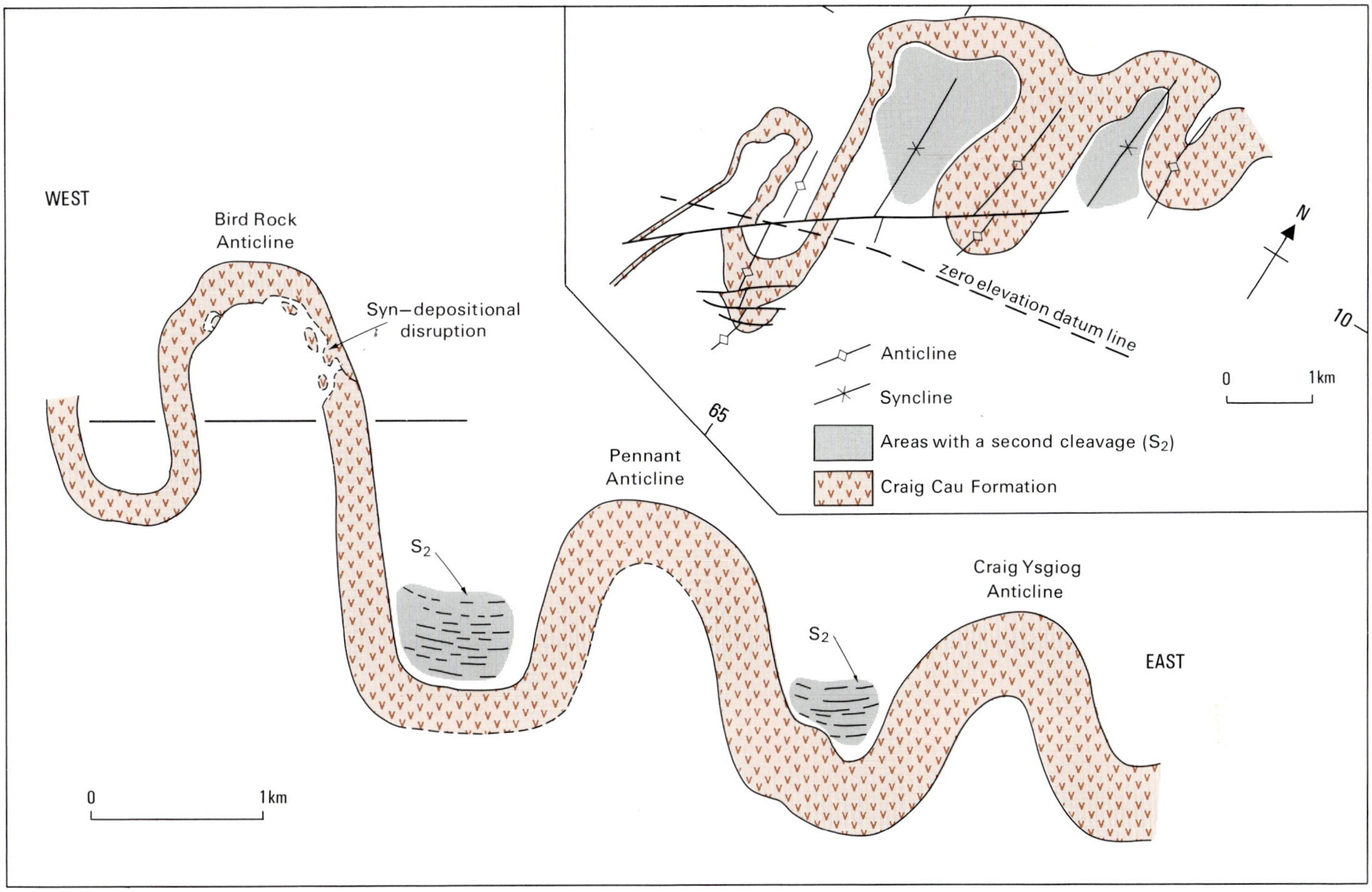

Figure 30 A profile of the Craig Cau folds constructed using the techniques of Wilson (1967) (modified from Pratt and Fitches, 1993). The distribution and orientation of the S_2 cleavage is also indicated.

but it transects their axial planes and there is no obvious inter-relationship.

Transection

The S_1 cleavage commonly lies clockwise of the accompanying folds. Transection angles, the minimum angle between the fold axis and the mean cleavage (Δ of Borradaile, 1978), are mainly around 10°, but, where fold axial planes strike north–south, are locally up to 26° (Figure 29). The largest angles occur around the folds of the Craig Cau Formation, particularly the Bird Rock Anticline, and Llandovery strata. Rare folds with east-north-east-striking axial planes, as north of Tal-y-llyn, are transected in an anticlockwise direction.

Transected folds, which are widespread in the Welsh Basin and Lake District (e.g. Soper et al., 1987; Woodcock et al., 1988; Woodcock, 1990), are interpreted to reflect anticlockwise rotation of folds, prior to cleavage development, during oblique compression (transpression). Competent beds, such as the Craig Cau Formation, which buckled after very little shortening, suffered the greatest rotation and developed the highest transection angles (Pratt and Fitches, 1993). Competency contrasts were the most important factor, but the forced development of folds above reactivated faults may have been important locally.

Faults

Faults are a prominent feature of the district and can be subdivided into three main groups based on their orientation: north-east–south-west (Tal-y-llyn, Bala, Dysynni, Ceunant), north-south (Llanegryn, Tonfanau), and east-north-east–west-south-west (Pennal, Cwm Sylwi). The main displacements on all the faults postdates the S_1 cleavage, although some may have had a pre-Caledonian expression. In addition, there are variably oriented faults which were clearly synchronous with end-Caledonian folding, such as the Forge and Gelli-goch 'Overthrusts', bedding-parallel faults within the black mudstones of the Nod Glas, Clogau and Dolgellau formations, and faults within fold hinges (Corris Fault).

North-east–south-west faults

The Tal-y-llyn Fault and the other elements of the Bala Lineament (Fitches and Campbell, 1987), are commonly considered to have affected Palaeozoic sedimentation (Bassett et al., 1966; Rast, 1969; James and James, 1969;

Smith, 1987b; Cope et al., 1992). However, within the district there is little evidence of syndepositional activity. For most of its length the fault lies within the outcrop of the Ceiswyn Formation and only in the north-east, where it joins the Bala Fault, can formations be compared across the fault. Even there, the minor thickness and facies changes within the Aran Volcanic Group cannot be attributed easily to syndepositional activity along the fault.

To the south-west of the district, the Tal-y-llyn Fault defines the edge of the Mesozoic/Tertiary basin in Cardigan Bay (Bullerwell and McQuillin, 1969; Dobson and Whittington, 1987). Between the coast and Bryncrug [6100 0220], it lies at the southern edge of the drift cover and its exact position is uncertain. From Bryncrug the fault lies along the valley to the head of Tal-y-llyn Pass where it joins with the Bala Fault and is deflected north-north-east (Figure 28).

The Tal-y-llyn Fault is probably a zone of anastomosing faults rather than a single fault. In the valley between Bryncrug and Dolgoch [6485 0470], gullies and linear knolls of faulted and slickensided mudstone, aligned parallel to the valley, are interpreted to reflect several faults. The fault zone is relatively well exposed where the valley narrows at Abergynolwyn [6720 0640] and Tal-y-llyn Pass. In a quarry [7530 1345] at the head of Tal-y-llyn Pass, tuffs of the Craig Cau Formation are crossed by numerous fault crush zones and in another quarry [7558 1383] mudstones with slickensided minor faults and kink bands are exposed.

The direction of downthrow on the Tal-y-llyn Fault is variable and the apparent downthrow is seldom more than 1 km. Between Cross Foxes and Abergynolwyn, it downthrows mainly to the north. At Tal-y-llyn Pass the Aran Volcanic Group is downthrown by 550 m and, near Abergynolwyn [6645 0700], repetition of the Craig Hengae Member indicates a vertical displacement of about 600 m. However, between Dolgoch and the Llanegryn Fault the fault downthrows to the south by about 1 km. West of the Llanegryn Fault, it probably downthrows 5–7 km to the north. Some faults within the zone, or nearby, downthrow in the opposite direction and effectively negate the net displacement on the fault system. Less than 1 km north of Tal-y-llyn [7154 0986], for example several north-east–south-west-striking faults downthrow the Aran Volcanic Group by 600–900 m to the south, and cancel out the downthrow on the Tal-y-llyn Fault. In this area the fault zone has the appearance of a foundered flower structure.

The configuration of outcrops and folds around the Tal-y-llyn Fault was probably achieved by both dip-slip and strike-slip displacement (Jehu, 1926; Cox, 1925). Axial-plane traces of some major folds are offset by the fault, and marked differences in the orientation of folds and cleavage across it cannot be explained solely by dip-slip movement. The Dol-Ithel and Mynydd y Llyn anticlines, which occur north and south of the fault respectively, were probably once continuous prior to a dextral strike-slip displacement of 3–4 km (Figure 28; Pratt, 1992). Likewise, fold/cleavage arcuations around the Dol-Ithel Anticline, north of the fault, and south-east of Abergynolwyn, south of the fault, may represent a single arc which was subsequently offset.

The Tal-y-llyn Fault caused little ductile deformation, or large-scale drag folding, and there is little systematic rotation of folds or cleavage into the fault zone. However, fault movements did create post-S_1 structures such as crenulation cleavages, conjugate kink bands and slickensided minor faults (Plate 15b; Bracegirdle, 1974). Their cross-cutting relationships suggest an older dextral strike-slip and younger dip-slip displacement (Pratt, 1992). The earlier structures comprise vertical, mainly east–west kink bands, in which the S_1 cleavage deviates sinistrally, and vertical north-east-striking minor faults with shallowly plunging slickensides. Stepped quartz fibres on the latter consistently indicate dextral strike-slip movement. Numerous kink bands and slickensided faults are well exposed at Dolgoch [6496 0447] and north-east of Abergynolwyn [6815 0775].

Neither the Dysynni nor the Ceunant faults have associated small-scale structures and there is little evidence of strike-slip movement. Both have an apparent downthrow to the north; the downthrow on the Dysynni Fault is probably similar to the two related north-east–south-west faults that downthrow the Aran Volcanic Group at Bird Rock [6458 0658], by 280 and 150 m. The hanging wall and footwall of the faults contain numerous, nearly horizontal quartz veins up to 2 m thick [6468 0660] which indicate dip-slip rather than strike-slip movement. South-west of Dolgellau, poor exposure and complex igneous intrusions prevent accurate estimates of displacement on the Ceunant Fault, but a large downthrow to the north, of approximately 1 km, is indicated by the offset of the Ffestiniog Flags Formation.

North–south faults

The Mochras, Llanegryn and Tonfanau faults define the east margin of the Mesozoic/Tertiary basin of Cardigan Bay (Figure 28; Dobson and Whittington, 1987). The Llanegryn Fault downthrows 5–7 km to the west and is probably the extension of the Mochras Fault. It can be traced from Llwyngwril to the edge of the drift cover at Peniarth Lodge [6050 0567] and it terminates against the Tal-y-llyn Fault. The fault zone is only exposed near Llanegryn [6034 0687] where it comprises a quartz-veined fault breccia of mudstone fragments. The Tonfanau Fault [5598 0370] lies mainly offshore, but at Tonfanau a borehole intersected a possible faulted contact between Tertiary sediments and Lower Palaeozoic mudstones (Dobson et al., 1973).

Geophysical evidence from the Mochras Fault indicates a mainly post-Jurassic and pre-Oligocene westerly downthrow of 2–6 km (Allen and Jackson, 1985) and the main movements on the Llanegryn and Tonfanau faults were probably contemporaneous. The evidence of Lower Palaeozoic, syndepositional movement on the Llanegryn Fault, on the basis of the pronounced thickness changes (Jones, 1933), is ambiguous and the changes may simply be the result of juxtaposition, by dip-slip, of depositional sites that were originally widely separated.

East-north-east–west-south-west faults

In the south of the district, a zone of east-north-east-striking faults, the 'transverse' faults of Cave and Hains

(1986), occurs between Happy Valley and Machynlleth. In the east, the zone is represented solely by the Pennal Fault, but, westwards, it becomes progressively broader and the faults form prominent depressions. Displacements, mostly dip-slip, are small and wallrock deformation is confined to post-S_1 brecciation and formation of kink bands. East of Machynlleth the Pennal Fault lies along the Dyfi Valley and downthrows some 150–200 m to the north. At Garth [7578 0166], an east–west branch downthrows by 60–70 m to the north and displaces the Forge Overthrust.

The Cwm Sylwi Fault appears to lie *en échelon* to the Pennal Fault (Figure 28) and is surrounded by small north-north-west-striking and north-west-striking faults, some of which are mineralised. A mineralised fault at Melyn Mine [SN 6178 9928] is marked by a 1 m-thick, quartz veined, fault breccia with vertical slickensides on the hanging wall. The Cwm Sylwi Fault passes south-westward, just off the south margin of the sheet, into numerous east-north-east-striking faults on Foel Caethle [SN 6040 9860] and north of Aberdovey (Cave and Hains, 1986).

FAULTS DEVELOPED DURING FOLDING

Some faults are closely associated with the folds developed during the end-Caledonian deformation. Faults in the hinges of large- and small-scale folds, particularly tight folds, may completely cut out fold limbs or synclinal hinges. They are generally less continuous than other faults in the district and include the Forge and Gelli-goch 'Overthrusts' (Cave and Hains, 1986) and axial faults within the Craig Ysgiog, Dol-Ithel and Corris anticlines. Some of these faults appear to develop as listric faults from the black mudstone formations, particularly the Nod Glas Formation. The strata above these ductile formations are commonly strongly faulted and folded compared with those below. It is suggested that localised *décollement* caused the overlying beds to deform independently of those below (Jones and Pugh, 1935; Cave and Hains, 1986; Martin et al., 1981; Pratt, 1991). The Ashgill strata at Corris are folded into a complex faulted anticline, but the underlying Nod Glas Formation is downwarped beneath the anticline, and associated reverse faults (with displacements of up to 450 m) appear to root into it (Figure 28). Similarly, the complementary fold, the Dyfi Syncline, terminates against the Nod Glas Formation. Furthermore, at Aberllefenni Quarry [7682 1010] the Nod Glas Formation strikes constantly north-east–south-west, but the overlying Ashgill strata are folded into a box fold approximately 1.5 km across. At the quarry, an east-north-east-striking reverse fault, possibly developed from the Nod Glas Formation, offsets the Narrow Vein Formation by 400 m but cannot be traced far into the Broad Vein Formation.

The black mudstones of the Nod Glas Formation are disrupted by bedding-parallel faults, kink bands, quartz veins, S_2 crenulation cleavages, folds of the S_1 cleavage and breccias. At Cwm Ratgoed [7739 1162], a 500 m strike section exposes two sets of faults, one of which lies parallel to bedding, and the other, to S_1 (Figure 31). The bedding-parallel faults are more persistent and cause more intense drag folding and brecciation of the wall-rocks. Reverse, and minor sinistral strike-slip, displacements are indicated by slickensides and drag folds of the S_1 cleavage, but the amount of displacement cannot be determined. Similar small-scale structures occur at Bryn-Eglwys Quarry [6923 0581], Craig Hengae [7593 1025] and Corris Uchaf [7445 0940]. At Corris the formation is repeated by a nearly bedding-parallel fault and is also highly disrupted. Quartz veins parallel to the S_1 cleavage are contorted and folded and, like those at Cwm Ratgoed, have an axial planar S_2 cleavage. At both localities, the drag folds of S_1 indicate that the faults continued to be active in the later stages of, or after, the end-Caledonian deformation.

The inliers of Ordovician rocks around Machynlleth are partly controlled by north-north-east-striking, west-dipping, reverse faults: the Forge and Gelli-goch 'Overthrusts', each with displacements of up to 500 m (Cave and Hains, 1986). The Ordovician strata crop out in the cores of large-scale, hanging-wall anticlines. In the foot-walls, bedding is commonly vertical or overturned. The geometry of the faults and the folds suggests that they are broadly contemporaneous (Jones and Pugh, 1916; Cave and Hains, 1986). In places, the faults dip at less than 45° and appear to be listric, possibly rooted in the Nod Glas Formation.

In the wavecut platform at Llwyngwril [5980 1095], a nearly bedding-parallel fault juxtaposes the Gamlan and Clogau formations and is marked by a gently folded, zoned vein of quartz and carbonate with intercalated leaves of mudstone (Figure 32). The folds of the fault plunge south at about 50°, some 20° more than the regional dip, and are more open than those in the adjacent mudstones. Minor listric reverse faults within the hinges of small-scale folds root into the fault as in a *décollement*, whilst others displace it slightly.

In the Dolgellau and Dol-cyn-afon formations, between Arthog and Friog, reverse faults strike mainly parallel to the local folds and S_1 cleavage and are either bedding-parallel or dip more steeply south-east. The faults are associated with considerable ductile deformation, including tight hanging-wall anticlines, suggesting that faulting and folding were contemporaneous. For example, an adit at Friog [6222 1286] exposes a nearly bedding-parallel fault at the top of the south-east-dipping Ffestiniog Flags Formation in contact with highly sheared, north-west-dipping mudstones of the Dolgellau Member. Similarly, near Arthog [6253 1340] the contact between these formations is also a reverse fault, with 130 m displacement, and the strata beneath the fault are highly disrupted with steeply plunging minor folds.

Regional metamorphism

The Lower Palaeozoic rocks of the Welsh Basin suffered low-grade, mainly sub-greenschist, regional metamorphism during the end-Caledonian orogeny. White mica crystallinity was enhanced by metamorphism and has proved to be a sensitive indicator of metamorphic grade (e.g. Roberts and Merriman, 1985; Robinson and Bevins, 1986). Within the

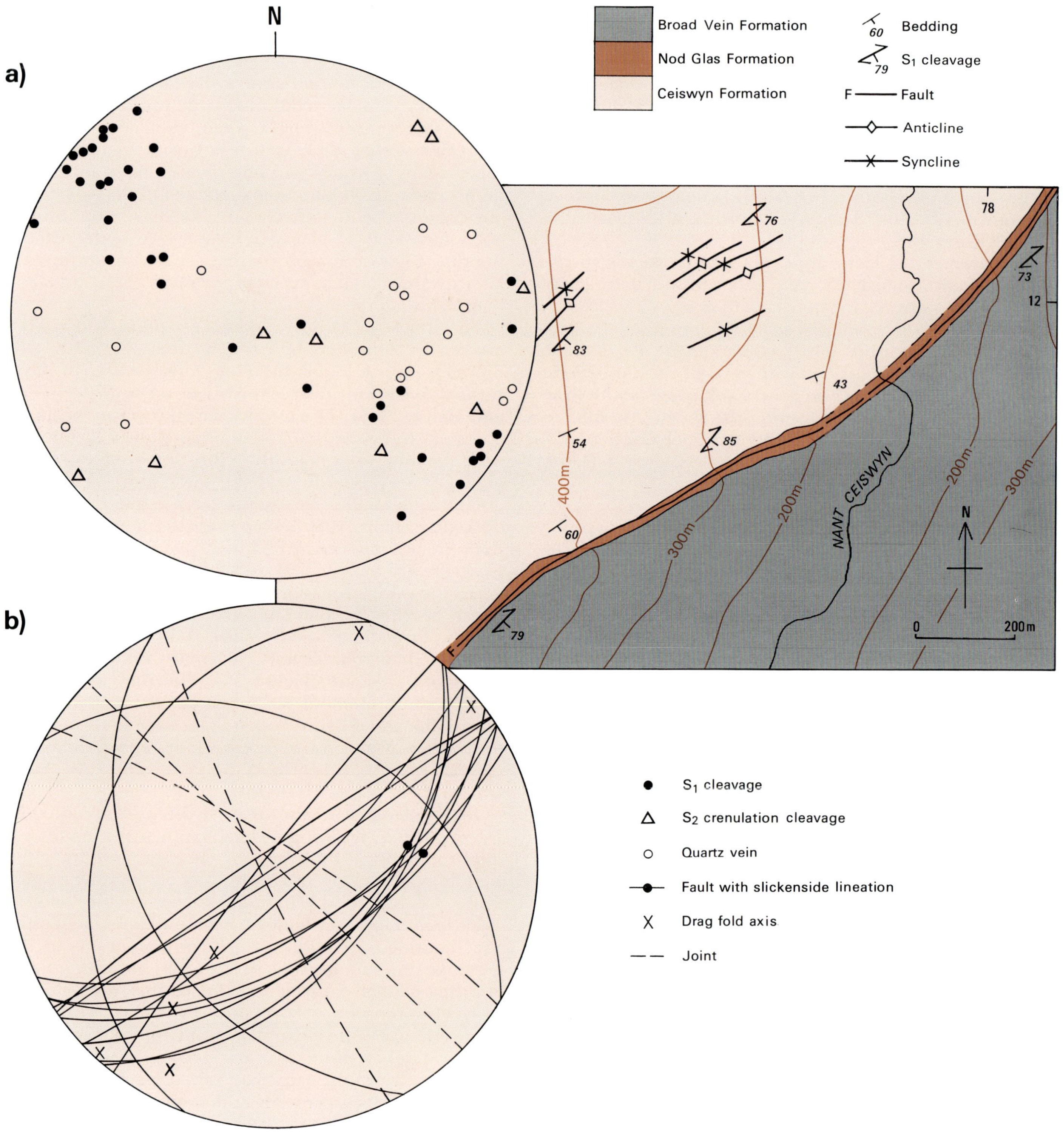

Figure 31 Structural data from the Nod Glas Formation at Ratgoed [7739 1162].
a) Stereogram showing poles to cleavage planes and quartz veins, b) Stereogram showing fault and joint planes (great circles) with slickenside lineations and drag fold axes.

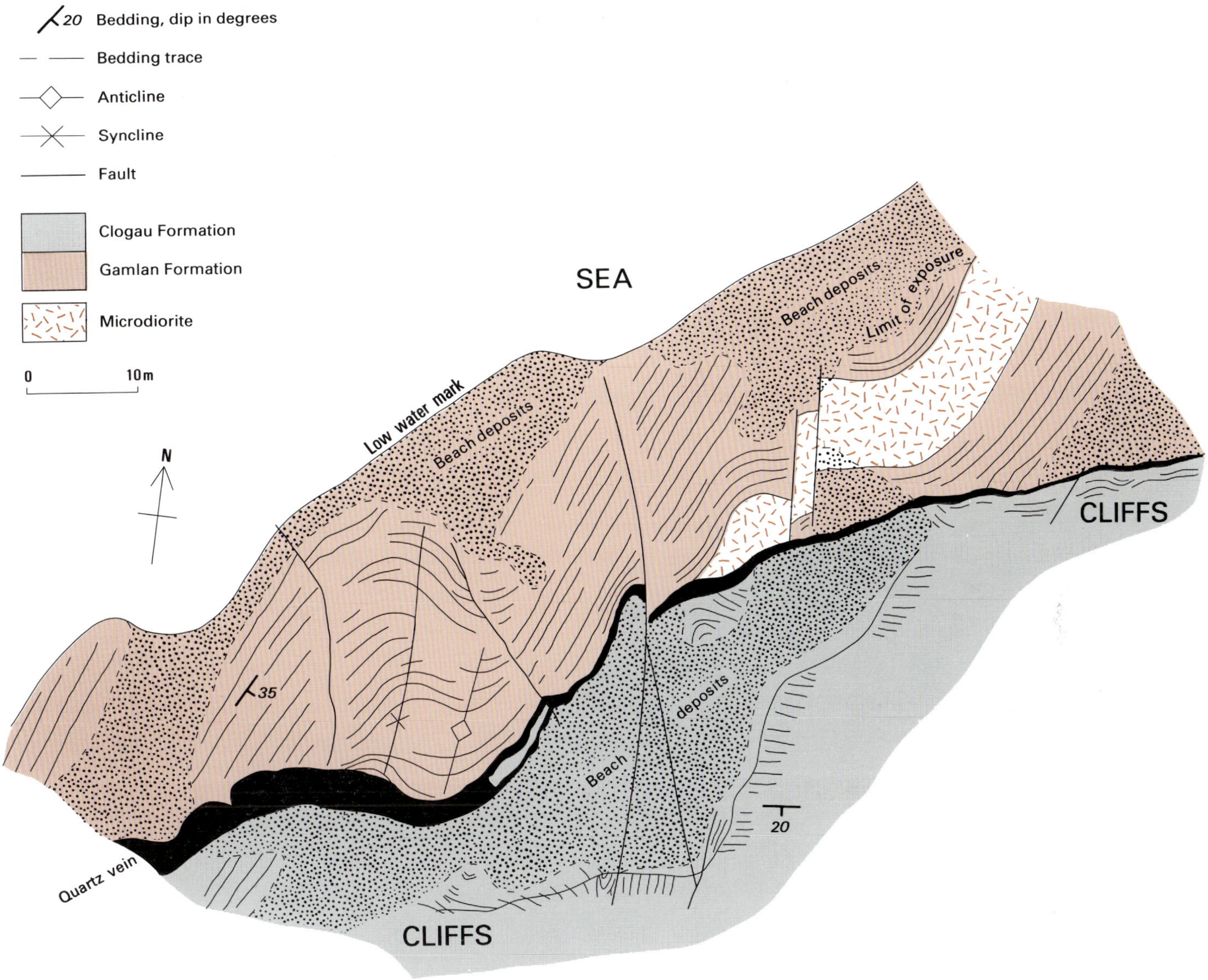

Figure 32 Plan of the wavecut platform at Llwyngwril [5980 1095] displaying a quartz-veined fault separating the Gamlan and Clogau formations. Also see Figure 6.

district, Roberts et al. (1991) recognised a decline in white mica crystallinity from older to younger rocks: the Cambrian strata are epizonal (low greenschist facies) whereas the Ordovician to Silurian strata are anchizonal (pumpellyite metabasite facies). However, this simple, depth of burial pattern was modified by superimposed inhomogeneous strain effects. For example, highly strained mudrocks adjacent to the steep limbs of folded competent layers, such as the Craig Cau Formation, display enhanced crystallinities. The Nod Glas, Broad Vein and Narrow Vein formations are also of higher grade than the underlying Ceiswyn Formation. This was ascribed by Roberts et al. (1991) to increased strain above a *décollement* within the Nod Glas Formation. They also noted that there is no anomaly or pattern of crystallinity about the Tal-y-llyn Fault or marked change across it (cf. Robinson and Bevins, 1986), although the white mica has retrogressed to smectite in nearby kinked and sheared mudrocks.

Many of the iron-rich rocks contain stilpnomelane. Pressure shadows around opaque ore grains within some acid tuffs comprise concentric layers of platy and fibrous stilpnomelane implying development during deformation (E67087, E67091). Stilpnomelane sprays are also developed on iron oxide residues along S_1 pressure solution seams in the mudstones. Porphyroblasts of chloritoid, a mineral commonly considered to be confined to the greenschist facies, are irregularly distributed within high anchizonal rocks of the Ty'r Gawen Mudstone Formation and the basal mudstones of the Ceiswyn Formation. The idiomorphic crystals, 30–100 μm in length, are mostly free of inclusions and their long axes lie nearly parallel to the S_1 cleavage. A minority of porphyroblasts are oriented at a large angle to S_1 and deflect the S_1 cleavage. Chloritoid growth was syntectonic, rather than pretectonic, but it probably extended beyond the development of the S_1 cleavage (cf. Prior, 1987). The main cause of chloritoid growth is poorly understood since neither the iron/manganese content of the host rocks nor the metamorphic grade appears to be controlling factor. Its restriction to deeper structural levels may indicate that its growth is pressure dependent.

The metamorphosed basic igneous rocks mostly comprise 'unbuffered' secondary assemblages of chlorite, actinolite, epidote, calcite and leucoxene and are of little use in determining pressure/temperature conditions (Liou et al., 1985; Cho et al., 1986). However, less-altered, buffered, assemblages of uralitic amphibole (actinolite), chlorite, pumpellyite and clinozoisite do occur in the centres of some intrusions (E67072). The absence of prehnite implies that pumpellyite-actinolite facies conditions were attained. The P/T grid of Liou et al. (1985) implies conditions of approximately 325° and 2.25 kilobars.

SEVEN

Geophysics

The Cadair Idris sheet lies on the boundary between the magnetically 'quiet' region associated with mid-Wales and the highly anomalous areas of the Harlech Dome and north-west Wales. This change corresponds broadly with the top of the Aran Volcanic Group and the Tal-y-llyn Fault. However, there is little expression of the Tal-y-llyn Fault in the gravity anomaly data, suggesting that there are no marked density contrasts across it. The gravity field is strongly influenced, both by changes at depth, possibly to middle or base crustal levels, which give rise to the regional increase in anomaly values westwards towards the Irish Sea, and also by the presence of lower density Tertiary–Mesozoic basins offshore.

Gravity data

The broad form of the gravity anomalies was outlined by Powell (1956) and Griffiths and Gibb (1965). The subsequent national survey programme increased the number of gravity stations within the district to one per 1.5 km^2. Although there are only nine network-adjusted offshore stations, which are less accurate than those on land, adjacent marine coverage (Dobson et al., 1973) allows reliable estimates of the contours. All the gravity values are now referred to the National Gravity Reference Net of 1973 (Masson Smith et al., 1974) and the Gravity Reference System of 1967 (Woollard, 1979). Saturated density values for Lower Palaeozoic rocks from north-west Wales lie within the range 2.65–2.80 Mg/m^3, with the acid volcanic rocks giving the lower values and the pelitic rocks the higher values (Smith *in* Allen and Jackson, 1985). In calculating the gravity anomalies from the regional data set (Figure 33a), a density of 2.70 Mg/m^3 was adopted for the Bouguer correction.

Bouguer anomaly values are highest in a broad north-east–south-west-trending ridge, centred on Tywyn, which dominates the onshore gravity field (A in Figure 33a). This gravity anomaly high is the culmination of a regional increase from near zero in the Welsh Borderlands to more than 40 mGal, locally, off the west coast. The increasing trend continues into Cardigan Bay, but is suppressed by the younger sedimentary basins. The Jurassic and Tertiary sequences to the north-west of Barmouth are sufficiently less dense than the Lower Palaeozoic formations, with contrasts of about 0.20 Mg/m^3 and 0.45 Mg/m^3 respectively, for basin thicknesses of 2–4 km to explain the observed decrease in gravity anomaly. However, the highest gravity anomaly occurs some 5 km inland and additional sources of density contrast have to be considered, such as changes within the basement or Lower Palaeozoic sequence in the vicinity of the coast.

Because of the lack of any consistent correlation between estimated thicknesses of Lower Palaeozoic rocks and the regional gravity anomaly pattern, the westwards increase has been attributed to effects deep within the crust beneath the Irish Sea (Blundell et al., 1971). However, a westward thickening of the relatively dense mudstone sequence may also make some contribution (cf. Carruthers et al., 1992). The northerly reduction in gravity anomaly values, towards the Harlech Dome, which begins about 10 km south-west of Cadair Idris where the contours are orientated east-south-east–west-north-west, also suggests an influence from elements at higher levels within the crust.

Steep gravity gradients are apparent along the coast, particularly near the offshore extension of the Llanegryn Fault (Figure 28), and reflect the contrast between the Mesozoic–Tertiary sedimentary basins offshore, and the Cambro-Ordovician strata onshore (B in Figure 33a). The magnitude of the gravity low is locally exaggerated because of the lack of marine data, as north-west of Barmouth, but the gradient across the Mochras Fault is well defined on land. Further inland, the relatively small density contrasts within the Lower Palaeozoic sequence inhibit clear definition of faults. Consequently, the Tal-y-llyn Fault is generally undetectable, despite the large downthrow between Bryncrug and Tywyn, and, although the Llanegryn Fault coincides with the north-east margin of the Tywyn high, it has little direct expression.

The residual gravity anomaly map of the district (Figure 33b), which suppresses the responses from deeper, mid-crustal sources, is dominated by arcuate, north-east–south-west to north–south trends which bear little relationship to the disposition of outcrops, but probably reflect the structural grain of the basement. By comparison with the Bouguer gravity anomaly map (Figure 33a), the Tywyn gravity high is more sharply defined and displaced north-west by a few kilometres. It is also distinctly arcuate and can be traced northwards into the Barmouth peninsula. Again, the Llanegryn Fault has little affect on the shape of the gravity high, but the ridge displays lower values, and is deflected to north-south, across the Mawddach estuary. Inland, a broad, arcuate zone of low gravity and low gradients is broadly parallel to the Tywyn–Barmouth gravity high. It is complicated by a local gravity anomaly high on the north flank of Cadair Idris (B in Figure 33b), around Mynydd-y-gader, which coincides with the thickly developed intrusive dolerite and extrusive basalt within the Aran Volcanic Group. To the south-east, a complementary low between Tal-y-llyn and Mynydd Moel (C in Figure 33b) corresponds to an area in which microgranite and/or the Craig Cau Formation crop out, or are close to the surface. South of Abergynolwyn (A in Figure 33b), there is further evidence of a correlation with local structures, but the gravity anomaly variations are of small magnitude and are not well defined.

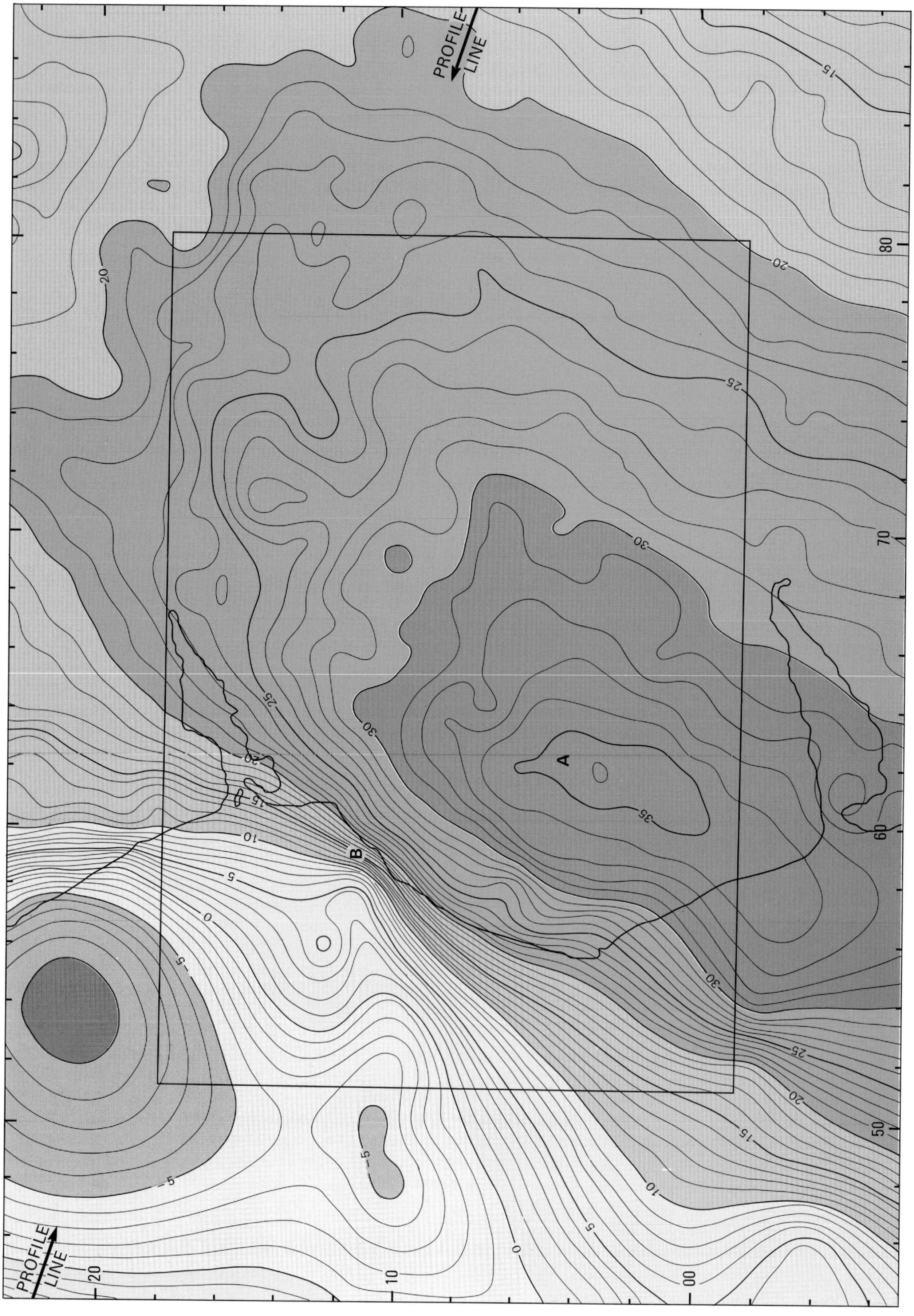

Figure 33a Regional gravity anomaly map of the Cadair Idris district and surrounding area. Reduction density 2.70 Mg/m^3; contour interval 1 mGal; colour shading interval 5 mGal.

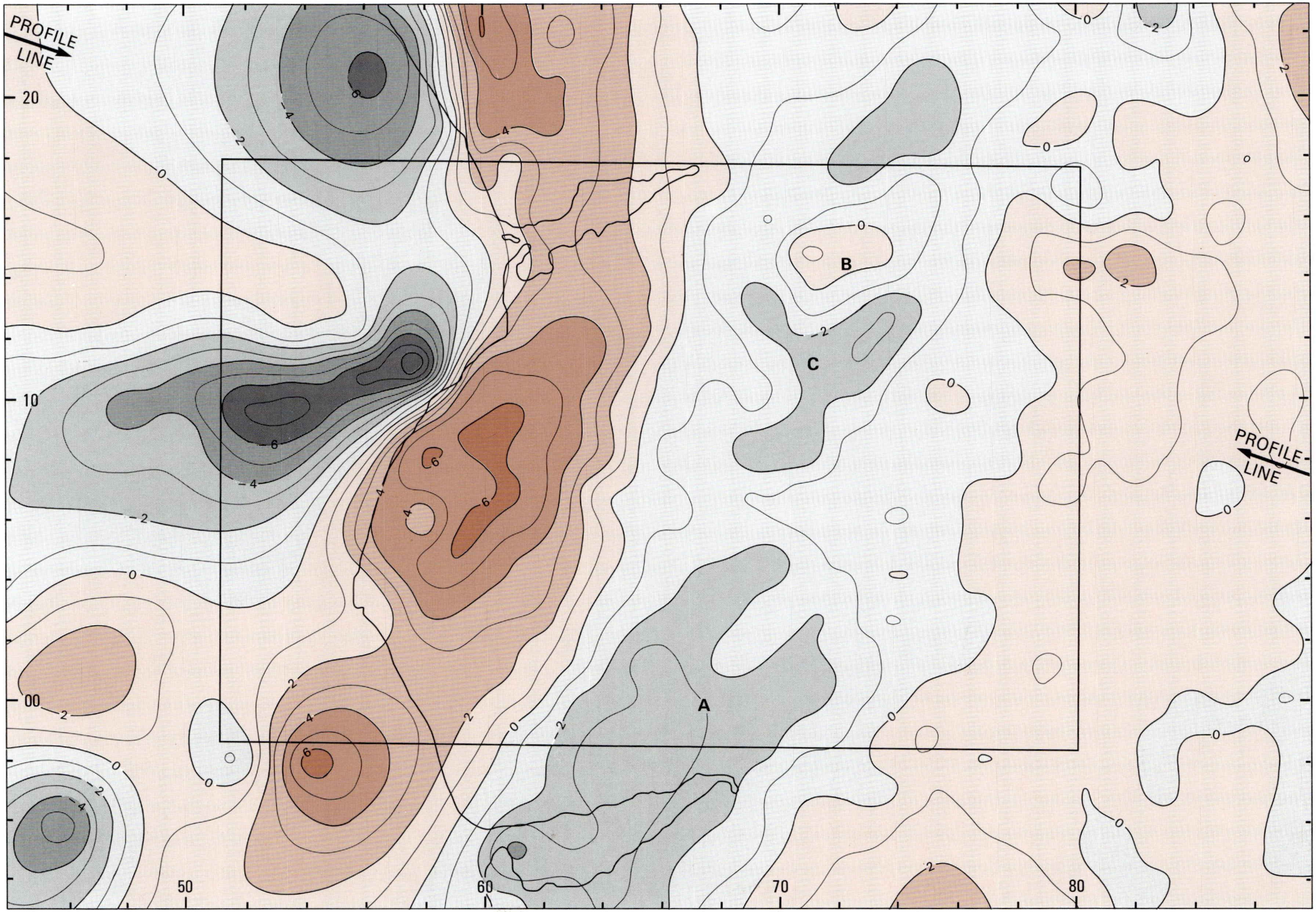

Figure 33b Residual Bouguer gravity anomaly map of the Cadair Idris district and surrounding area (after applying a 20 km wavelength high-pass filter to the regional data set). Contour interval 1 mGal; colour shading interval 2 mGal.

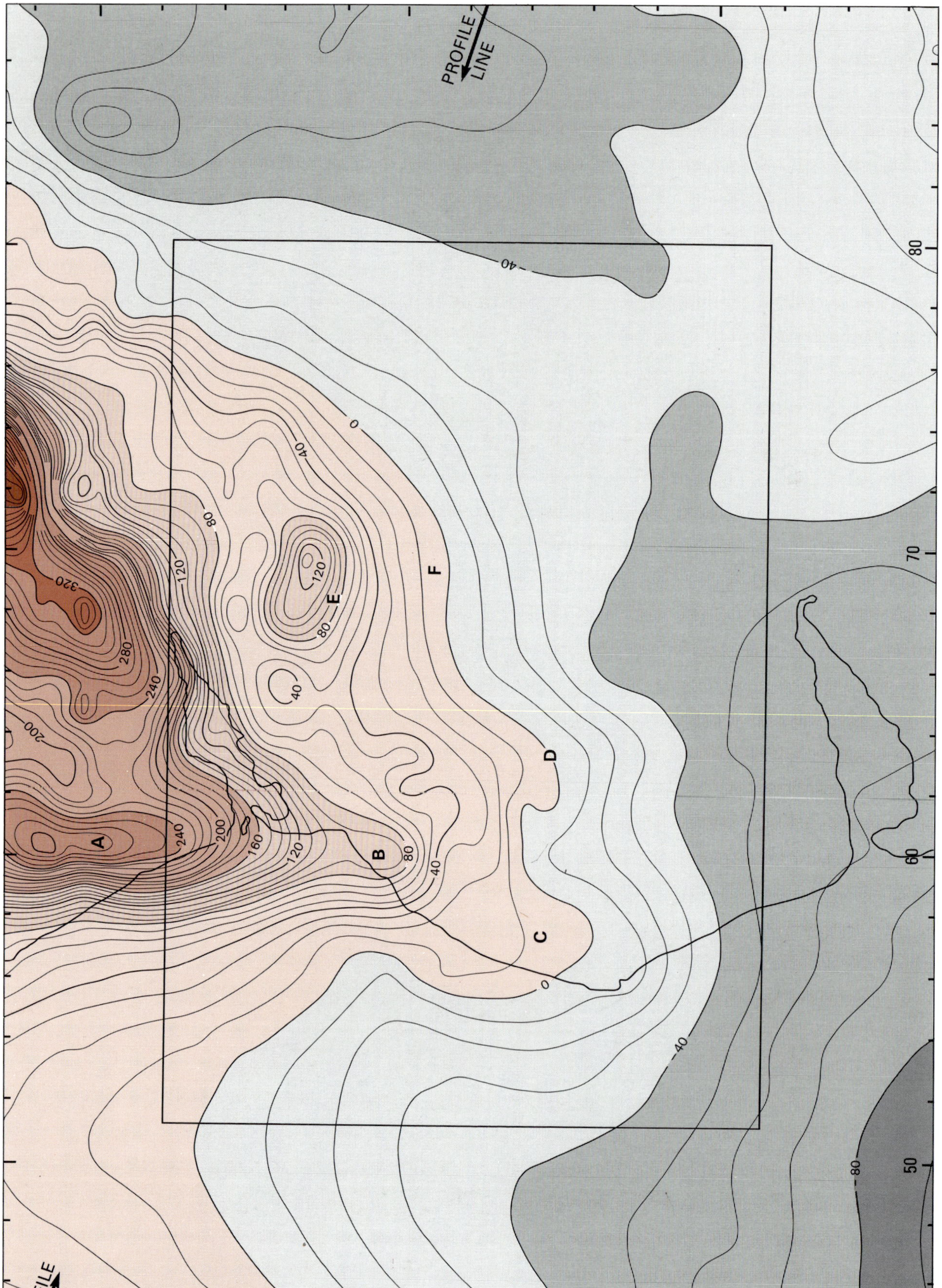

Figure 33c Regional aeromagnetic anomaly map of the Cadair Idris district and surrounding area. Flying height 305 m; contour interval 10 nT; colour shading interval 40 nT.

Detailed gravity surveys in the lower reaches of Afon Dysynni, from Peniarth to the coast, with a station interval of about 400 m, determined a residual gravity anomaly with an amplitude of 2 mGal and a width of about 1 km (Blundell et al., 1969). The anomaly approximates to the course of the Afon Dysynni and was interpreted as a buried valley with a depth to bedrock of up to 150 m. A mean density of approximately 2.20 Mg/m^3, close to that of till, was adopted for the valley fill.

Magnetic data

An aeromagnetic survey across the district was flown as part of the national coverage. Data were collected along north–south lines, 2 km apart, and east–west lines, about 10 km apart, at a nominal ground clearance of 305 m. The total magnetic field anomalies (Figure 33c) are referred to the planar reference field which was derived at the time of the initial reduction of the full UK data set (Geological Survey of Great Britain, 1965). North of the Mawddach estuary, a more detailed survey was flown along north-west–south-east lines, 200 m apart, at a ground clearance of approximately 45 m (Allen et al., 1979).

Magnetically, the north part of the district resembles the Harlech Dome, with discrete anomalies of near-surface origin superimposed on an elevated background. The short-wavelength anomalies were clearly defined in the low-level aeromagnetic survey (Allen et al., 1979), particularly north-east of Barmouth, where a series of short-wavelength anomalies, not apparent in the standard map presentation of Figure 33c, clearly originated close to the surface. While the Harlech Dome is underlain by magnetic 'basement' at relatively shallow depths, there is a thick cover of nonmagnetic rocks in the south of the Cadair Idris district, as shown by the marked reduction in anomaly gradients.

Within the Cambrian sequence, magnetic beds occur in the Rhinog Formation and the underlying Dolwen Formation (Allen and Jackson, 1985). Pronounced anomalies also occur above the outcrop of the Maentwrog Formation, but, where measured, these rocks are generally less magnetic. Specific anomalies can be related to igneous intrusions, but generally their distribution and inconsistent level of magnetisation suggest their overall contribution is small. Despite the locally high magnetisation within the Cambrian strata, their volume is insufficient to account entirely for the background levels over a wide area in the north of the district and a significant contribution from the Precambrian basement is necessary.

The highest regional values occur near Barmouth, on a north–south-trending ridge (A in Figure 33c). The steep gradient on the west side of this anomaly can be attributed to the termination of magnetic Cambrian strata at the Mochras Fault. South of the Mawddach estuary, outside the Cambrian outcrop, the north–south-trending anomalies continue but are more subdued, probably because the depth to the magnetic Precambrian basement is greater. Near Llwyngwril, the Barmouth high coincides with the outcrop of Maentwrog Formation (B in Figure 33c) and to the south this ridge divides into two weak anomalies, C and D, on either side of the Llanegryn Fault.

Apart from a distinctive magnetic anomaly high (E in Figure 33c), immediately west of Cadair Idris, the Aran Volcanic Group rocks do not appear to be strongly magnetic. The pronounced topography may distort this anomaly and, consequently, the quantitative interpretations are less reliable. It probably reflects high-level elements within the crust since the anomaly is defined sharply and is centred on the thickest development of the Cadair Idris microgranite. However, it is not associated with a comparable gravity anomaly.

The distinct reduction in magnetic values in the south of the district (F in Figure 33c), is probably due to the increased depth of burial of the magnetic basement. No magnetic strata are distinguished within the Ordovician–Silurian sedimentary rocks and any major underlying source must lie at depths exceeding 4 km.

Seismic data

Offshore, deep seismic profiles provide a valuable insight into the structure of Cardigan Bay. North-north-west of Tonfanau, measurements from four reversed refraction lines, with a total length of about 15 km, identified three seismic layers (Blundell et al., 1964): an upper layer, not present onshore, 80 m thick and with a velocity of 1.5 km/s, an intermediate layer, 150 m to 600 m thick, of 1.8–2.3 km/s, and an underlying layer with a velocity of 3–3.8 km/s. In addition, there was evidence of a faster layer, more than 4.5 km/s, below 1 km depth. The geological attribution of these layers remains uncertain, particularly for the lowest which is provisionally designated as Ordovician.

An offshore seismic reflection survey (Bullerwell and McQuillin, 1969) indicated a thick sequence of Mesozoic/Tertiary sedimentary rocks above Lower Palaeozoic rocks. Within the district, an east–west-oriented seismic line showed east-dipping strata which are apparently faulted out near the axis of a basin containing about 350 m of Tertiary, 2350 m of Jurassic and 1800 m of Permo-Triassic strata. Offshore, shallow reflection data distinguish the near-surface formations more clearly and assist in mapping the pre-Quaternary outcrop, such as the tongue of Tertiary rocks preserved against the Tonfanau Fault south of Barmouth Bay and intersected by the Tonfanau borehole (Dobson et al., 1973).

In the lower Dysynni Valley, a shallow seismic refraction experiment distinguished bedrock velocities of 3.9–4.6 km/s, typical of a weathered and fractured competent rock, similar to those values determined from the Ceiswyn Formation near Tywyn (Blundell et al., 1969). Velocities of about 1.8 km/s were given by the unconsolidated valley fill and the depths to bedrock compared closely with those derived from the gravity data. A shallow intermediate layer, up to 25 m thick, with a velocity of 2.4 km/s, was also determined locally.

Two-dimensional models

Interpolated gravity and magnetic values were modelled along a west-north-west–east-north-east profile (Figure 34), broadly corresponding to the northern line of section

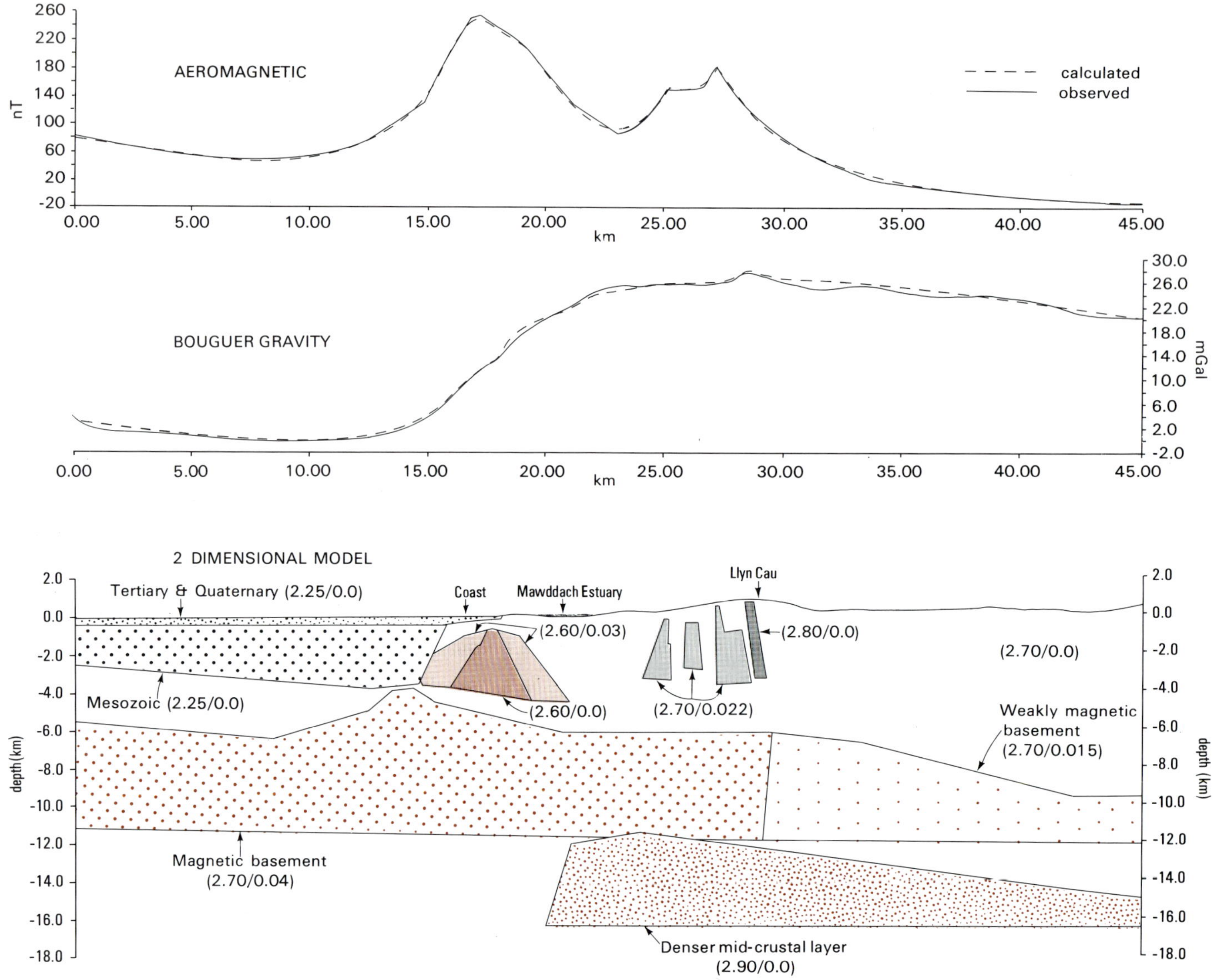

Figure 34 2D profile from [4400 2200] to [8800 0750] showing aeromagnetic and gravity anomaly data together with a schematic model. See Figure 33 for location. Polygon properties are given as density (Mg/m^3)/magnetic susceptibility (10^{-3} SI units).

on the 1:50 000 Geological Sheet (149). The aeromagnetic anomaly datum was adjusted down by 40–50 nanoTesla and the gravity anomaly background was attributed in part to crustal thinning of 1–2 km towards the Irish Sea. The resulting two-dimensional model indicates the possible form of the sources necessary to account for the anomalies.

The three major features of the geophysical interpretation are a mid-crustal layer of higher density, culminating near the coast, a magnetic, Precambrian basement and variations within the shallower Lower Palaeozoic strata and intrusions.

The depth of the offshore Mesozoic basin, which is constrained by seismic data, does not account fully for the decrease in gravity anomaly at the coast. Rocks of intermediate density (2.60 Mg/m^3) are included in the model in order to reproduce the observed gravity anomaly gradient and also contribute to the magnetic anomaly. They probably include Cambrian rocks although the density is unusually low. The polygons between 25 and 30 km represent the basic intrusions around Cadair Idris. The underlying magnetic basement has to be separated from the denser body at depth in order to match both sets of anomaly patterns.

EIGHT

Tertiary and Quaternary

Within the district, the only possible Tertiary deposits were encountered in the Tonfanau Borehole [5600 0376] (Institute of Geological Sciences, 1971, p. 123). The deposits, about 70 m thick, overlie Cambro-Ordovician strata at 107 m, and are overlain by 36 m of Quaternary clays, sands and gravels. They mainly comprise upward-fining sequences of clay, sand, conglomerates and breccias with some evidence of lignite. An abbreviated log of the borehole is given in Appendix 3. The clays contain no microfloras, and the microfaunas were nondiagnostic (British Geological Survey, Report PDL/72/19). However, lithological similarities with the Oligocene–early Miocene sequence of the Mochras Farm Borehole, interpreted as meandering river channel and floodplain deposits (O'Sullivan, 1979), suggest a similar age. At Tonfanau, geophysical investigations suggest a major coastal fault which downthrows to the west (Dr R J Whittington, personal communication, 1992). However, its precise position is uncertain since the contact in the borehole was gently inclined and 'breccias' between the Cambrian/Ordovician and Tertiary resembled weathered rock rather than fault breccia.

The topography of the district indicates that during Quaternary times it was subjected to intense glaciation. The features are probably the product of Late Devensian glaciation which were subsequently modified by periglacial activity and, in the high cwms, by a brief glaciation at the end of the Devensian. The district lay just to the south of the thickest ice accumulation in Wales, between Arenig Fawr and the Rhinog Mountains (Campbell and Bowen, 1990). This ice was entirely of local accumulation, about the high ground of north-west Wales and, during the major glacial episodes in Quaternary times, it obstructed the southerly movement of the larger, Irish Sea Ice Sheet about the north and west sides of Wales. The topographic grain of the district indicates that the main direction of ice movement was from the north-east to the south-west.

The most spectacular cwms and cwm lakes occur around the summit of Cadair Idris. Cwm Cau is probably the most impressive with precipitous confining walls rising close to the level of the summit (Figure 35). On the north edge of the escarpment, Cwm Gadair is a closely comparable feature although here the back wall is excavated into the well-jointed microgranite sheet. Eastwards along the escarpment, to Cwm Aran, the microgranite cliff was similarly scalloped by the ice, but with less spectacular cwm development. The south-facing Cwm Amarch on the south-west side of Cwm Cau, was excavated into the dip slope of the silty mudstones of the Ceiswyn Formation which resulted in less precipitous slopes. Away from the summit area, Cwm Cyri is probably the best-developed feature but steep, ice-excavated valleys occur throughout the district.

In the lower ground, the scouring effect of the base of the ice sheet is most graphically displayed where it affects a substrate comprised of alternating, markedly contrasting lithologies. Probably the best example of this in the district is the serrated profile of the terrain to the east of Llynnau Cregennen, which is underlain by volcanic rocks interbedded with silty mudstones of the Aran Volcanic Group. The resistant volcanic rocks form ridges, such as Pared y Cefnhir and Bryn Brith, and the intervening depressions mark the outcrop of less resistant, and easily eroded, cleaved, silty mudstones. Erosion produced more typical roche moutonnée forms where less well-bedded sequences were involved, especially where the outcrop comprised thick, uniform lithologies such as an igneous body. There are many examples of such features in the district and they are particularly well developed across the dolerite intrusions, as on the flank of Mynydd y Gader, and on thick basalts, as in the broad valley east of Llyn Cau.

The north-east-striking Tal-y-llyn, Dysynni and Gwernan valleys were all excavated along major faults. From Tal-y-llyn Pass, the view to the south-west is the most impressive aspect of the valley, across the classical glacial trough occupied by Tal-y-llyn. It is in the upper part of the valley that the tributary hanging valleys, such as Cwm Cau and Cwm Amarch, with their series of waterfalls, are most spectacular.

The drainage history of the district is complex and reflects the modification of a pre-glacial, north-west-trending drainage system, possibly developed by superimposed drainage through a cover of Mesozoic strata, by glacial excavation of north-east-trending valleys, such as the Tal-y-llyn, Dysynni and Gwernan (Figure 35). Many former north-west-trending channels can be traced across the Tal-y-llyn valley and are represented to the north by dry valleys, as at Abertrinant [6430 0500] (Figure 35). The most striking example of anomalous drainage occurs at Abergynolwyn, in the Tal-y-llyn valley, and was caused by a landslip [6710 0640]. There, the Afon Dysynni turns through almost 90° and cuts across the watershed between the Dysynni and Tal-y-llyn valleys, returning to a previous, north-west-trending, preglacial drainage channel.

The area has attracted a large number of geomorphologists over many years and, in this respect, the work of Dr E Watson should be particularly acknowledged. Recently the Cadair Idris massif has been designated, for its Quaternary features alone, as a possible Site of Special Scientific Interest (SSSI) and, to this end, its details have been compiled by Campbell and Bowen (1990).

Quaternary deposits

Throughout the district there is a wide range of drift deposits although clear designation of their distribution is

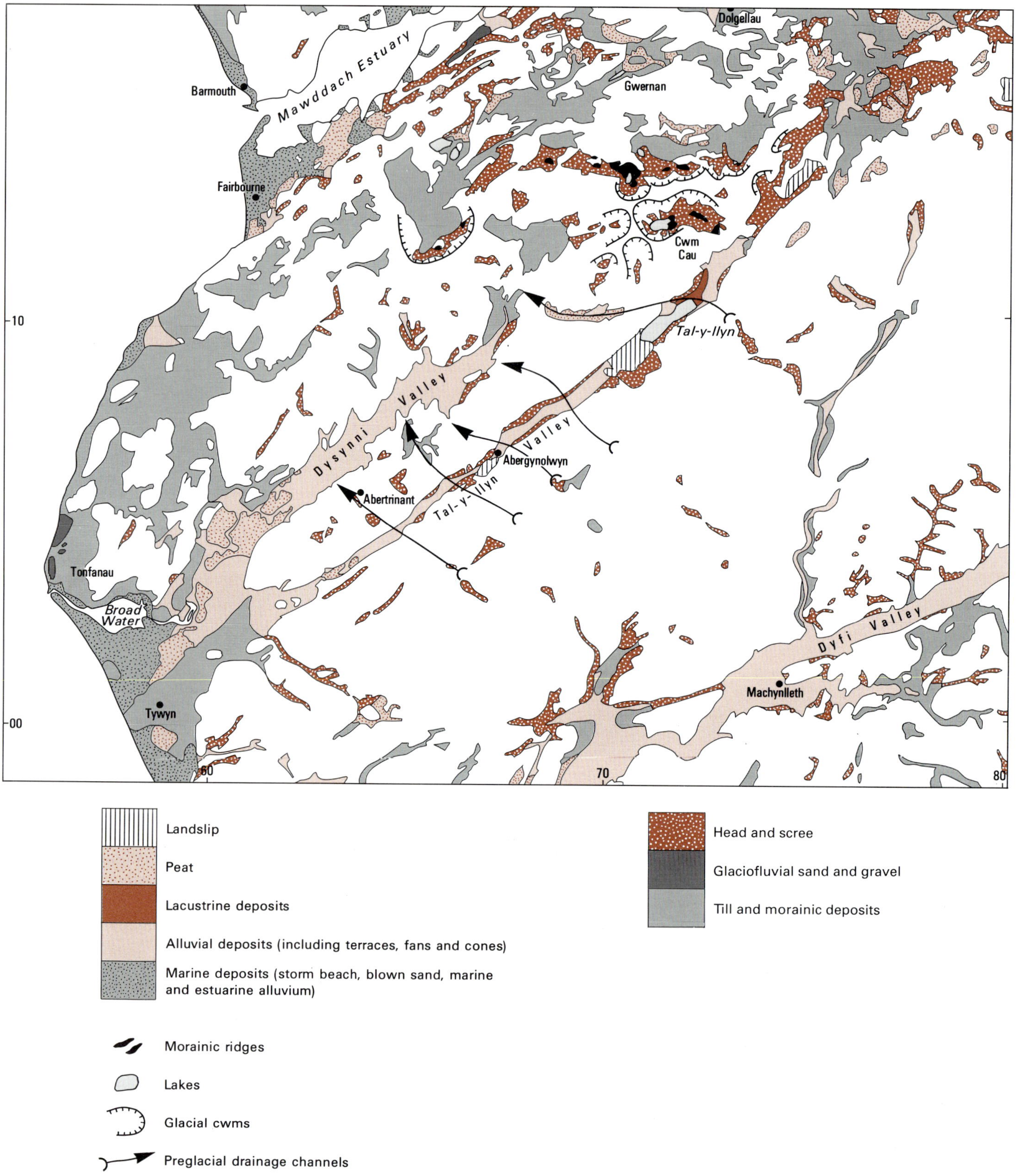

Figure 35 The major landforms and drift deposits of the Cadair Idris district.

commonly difficult because their boundaries are indistinct and they commonly grade into each other.

Hummocky glacial deposits (moraine)

In spite of extensive solifluction during postglacial times, there are many examples of well-formed moraines especially in the high cwms about the Cadair Idris summit, for example Llyn Aran (Plate 16a). Most are constructed of boulders and gravel of local lithologies, mainly igneous and volcanic rock. At the eastern edge of Llyn Cau, low moraine mounds, scattered with blocks, overlie the eroded rock lip and extend on to the lower edges of the confining slopes, especially those on the north side. Probably the largest and best-developed group of moraines are those at the north side of Llyn Gadair which were locally breached by drainage from the lake. They can be traced out of the confines of the cwm, suggesting that their distribution was partly controlled by the steep escarpment. Smaller, but well-developed, isolated moraines lie perched on the steep slope in the back wall of Cwm Gadair.

Away from the high cwms, there are larger, well-defined moraines that probably developed marginally to the main valley glaciers. Typical of these is the well-rounded feature [7300 1230] north of Minffordd in the Tal-y-llyn valley, and the well-featured terraces [6700 1385], on the steep slope, south-east of Llynnau Cregennen. Sarn y Bwlch, an east–west-trending ridge of boulders that extends offshore from Tonfanau, may also be a large moraine, developed at the confluence of the main outflows from the Mawddach and Dyfi/Dysynni estuaries.

Till (boulder clay)

Till deposited from the base of the ice sheet is widespread in valley floors in the north of the district, but the thickest accumulations occur along the coastal fringe between Tywyn and Llwyngwril, where they adhere to relatively steep slopes. The inland deposits are mainly those of the local Welsh Ice sheet and typically comprise grey-brown clay, superficially weathering to ochreous brown, with poorly sorted and locally derived gravel and boulders. An ill-defined layering, caused by alignment of clasts, is only rarely apparent. In general, the deposit closely reflects the substrate and is rich in small angular

Plate 16a Moraines around Llyn Aran [7345 1385].

Plate 16b Cross-bedded gravels in river terrace near Abergynolwyn [6841 0763]. Face is about 6 m high.

Plate 16c Glaciofluvial gravels overlain by till (A14581). Tonfanau [5617 0448].

mudstone fragments where it overlies cleaved silty mudstone, or it is characterised by isolated rounded boulders where it overlies blocky, jointed volcanic rocks. However there is a strong tendency for the clay fraction to decrease on the higher slopes, where the deposits comprise mainly clasts with little matrix.

The till in the coastal fringe displays a diverse assemblage of rounded pebbles and boulders, as south of Llwyngwril [5938 0777], where cobbles, up to 0.2 m across, include porphyritic and vesicular lava, tuff, microdiorite, microgranite, vein quartz, greywacke and dolerite. In the Tonfanau coastal section [5603 0354 to 5616 0442], the most distinctive clasts are blocks of fossiliferous, Jurassic limestone and semiconsolidated rafts of lignitic, Tertiary clay. There are no onshore sections of Jurassic within the district, but it has been recorded in the Mochras borehole to the north (Woodland, 1971), and offshore (Dobson and Whittington, 1987). The distinctive boulders and sandy fraction of this till suggest it was deposited by the Irish Sea Ice sheet. At Llwyngwril [5915 1025], a coastal section exposes up to 9 m of stiff brown clay with boulders and lenses of sand and gravel. At Tonfanau and Llwyngwril, the uppermost till commonly contains ice-wedge casts (Watson, 1967).

Drumlins are not common. Probably the best developed is the east–west mound on which Tywyn is sited. Its margins are clearly defined only locally, as near the hospital [5908 0045]. The drumlin deposit is relatively clay rich, and the banks of the cutting occupied by the Tal-y-llyn Railway are prone to subsidence and flows of clay after heavy rainfall. Temporary excavations of till in Tywyn cemetery [5953 0136] exposed 2.2 m of poorly sorted gravels with interstitial ochreous silty clay and few well-rounded boulders, up to 0.2 m in diameter, of acid volcanic rocks and sandstone.

In places, as near Waenfach, the upper surface of the till has been ferrocreted. A stream section [5903 0506] exposes gravelly clay, 1.25 m thick, overlain by a 0.2 m layer with a concentration of small boulders in an iron rich cement.

Glaciofluvial deposits

Glaciofluvial sand and gravels are thickest along the coastal fringe, where they commonly interdigitate with, or occupy channels on top of till. The best exposed sections occur in the sea cliffs at Tonfanau [5600 0360] (Plate 16c) and at Llwyngwril [5922 1025]. Typically the sand and gravel occurs in channels, up to 2 m deep, within the till. To the north-east of Tywyn [5968 0213], glaciofluvial deposits form a line of low mounds, up to 4 m above the level of the surrounding alluvial deposits, possibly representing a buried esker. The deposits [5975 0194], comprise well-rounded pebbles, 0.05–0.2 m in diameter, of acid volcanic rocks, sandstone, mudstone and quartz. North-east of Arthog, near Abergwynant [6721 1717], a pit exposes up to 5 m of interbedded sand and gravel. The bedding dips steeply (50°) east, possibly as a result of deformation by overiding ice.

Near Aberllefenni [7630 1070], glaciofluvial gravels, with a sparse clayey matrix, overlie alluvium and are overlain by head. To the south, a similar moundy terrace feature borders the alluvium for much of the length of the Dulas Valley. In Cwm Ratgoed, a track section near Ceiswyn [7778 1250] exposes 2 m of well-rounded, imbricated pebbles of mudstone and volcanic rock.

Scree and head

Screes are a distinctive feature in the high cwms of the Cadair Idris escarpment. Even though they are mainly the product of periglacial conditions they are still accumulating, facilitated by well-developed joint patterns.

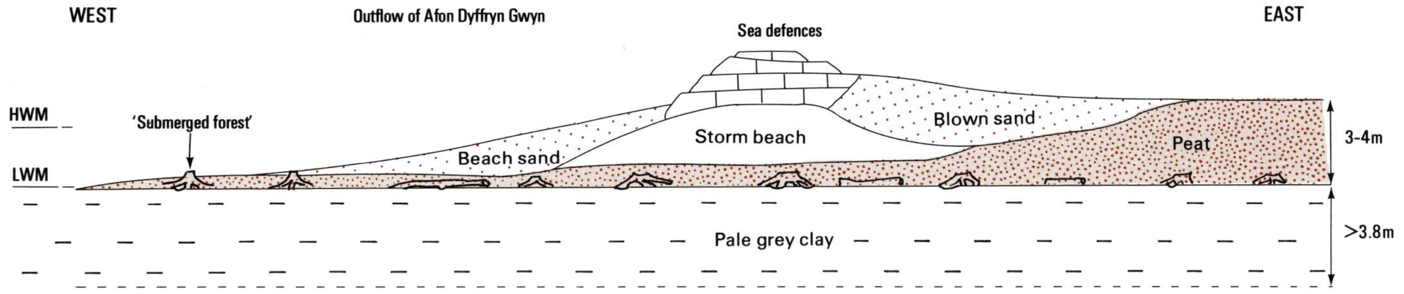

Figure 36 Cross-section of the storm beach and submerged forest at Tywyn [SN 5810 9970], as established by augering.

The deposits are most extensive beneath the microgranite, notably between Cyfrwy and Mynydd Moel, and comprise cobbles and boulders with a resting angle of 30–40°. Most of the scree slopes below crags of silty mudstones in the Tal-y-llyn valley [7130 0900] are stabilised by vegetation and thin soils, but a few, mainly those on steep slopes, are still active. Mudstone screes are characteristically stratified (*Grèzes litées*), with coarse and fine rock fragments, and were formed by freeze–thaw processes (Watson, 1965, 1968, 1977). Typical is a quarry in Tal-y-llyn Pass [7540 1365] which exposes 5 m of stratified gravel, with the uppermost metre displaying well-developed, ice-wedge, contraction fractures.

Head deposits are widespread, although only the more significant have been mapped. On the steeper slopes, particularly those which are vegetated, the division between scree and head deposits is not easy to distinguish as they grade into each other. Most vegetated slopes are underlain by a layer of soliflucted debris, especially those developed above well-cleaved sedimentary rocks. Typical are the deposits in the vicinity of Corris, as west of Godre Fynydd [7520 0972], which comprise crudely stratified, angular mudstone fragments within a sparse, ochreous, silty clay matrix. Locally, as in Cwm Celli [7590 0960], layers of peat are intercalated with the gravel deposits which, near Llanwrin [7767 0358], are up to 5 m thick.

Marine, shoreface, blown sand and beach deposits

A prominent storm beach fringes the coastline from Fairbourne, in the north, to Tywyn, in the south. It is best developed, 5 m high and 120 m wide, at the mouth of Afon Dysynni and in the Fairbourne spit. Several 'fossil' storm beaches, marking the progressive northward construction of the spit, can be distinguished between Friog and the mouth of Afon Mawddach, and around the mouth of the Afon Dysynni. The storm beaches mainly comprise diverse, well-rounded pebbles, 0.02–0.15 m in diameter. Near the mouth of Afon Dyffryn Gwyn the storm beach is lenticular in cross-section and it overlies peat; on the western, seaward side it is overlain by beach deposits and, to the east, by blown sand (Figure 36).

Around Tywyn, degraded and vegetated sand dunes form a tract up to 500 m wide behind the storm beach. They are best developed on the west side of the Broad Water [5680 0280] where they form arcuate ridges aligned parallel to the ancient storm beaches. The west side of Tywyn lies mainly on hummocky blown sand and temporary excavations have exposed at least 1.25 m of ochreous sand with thin fossil soil horizons.

Estuarine and marine alluvium are best developed within the Mawddach, Dysynni and Dyfi estuaries and a recent borehole on Fairbourne spit [6154 1486], begun on the storm beach, proved at least 48 m of marine and estuarine sands, cobbles and clays, without encountering solid rock. Because of land reclamation, the size of the Dysynni estuary has diminished considerably in the last few centuries. However, at some time in the recent past, estuarine alluvium was deposited at Craig yr Aderyn (Bird Rock), currently about 9 km inland, and was subsequently covered by the river alluvium. Both estuarine and marine alluvium comprise pale grey clay with rootlets, representing vegetated mudbanks, and brown sands, reflecting deposition within tidal channels, and are well exposed in small banks about the Broad Water and Afon Dysynni.

The well-drained land between the Broad Water and Tywyn is underlain by predominantly sandy, estuarine alluvium, which is exposed in drains [5876 0223]. To the east it overlies and interdigitates with peat, and to the south it is overlain by old sand dunes, forming the hummocky ground west of Morfa Camp [5810 0080].

Alluvial fans and fan deposits

Alluvial fans develop where a stream undergoes an abrupt change in gradient and, as a result, some of the largest fans lie at the confluence of vigorous, tributary mountain streams with the main valleys. There are numerous examples in the district, particularly in the Tal-y-llyn and Dyfi valleys. The fan at Pentre [7170 1030], on the north-west side of Tal-y-llyn, has a radius of 400 m and continues to grow. On the north-east side of the lake, a series of coalescing fans, locally deeply incised by subsequent drainage, has developed at the confluences of Ceunant Dol-ffanog [7300 1034], Nant Cildydd [7300 1133] and Nant Cadair [7278 1176] with the main valley.

The alluvial fans are formed of crudely stratified debris, the composition of which reflects the eroded substrate, and the shape of which reflects the distance of transport. Most typically they comprise subangular to rounded pebbles and boulders with impersistent lenses of gravelly and silty clay. Steep-sided fans comprise nonstratified boulders, whereas gentler fans resemble alluvium, with stratified gravels, sands and clays. Exposures are generally small and restricted to the current stream course and to the uppermost layers of the larger deposits, which are likely to be thick.

Fan deposits comprise gravel, sand and clay with irregular form and flat, or only gently inclined, tops. Some deposits have a similar origin to the alluvial fans, for instance the large deposit that spills from the mouth of the Tal-y-llyn valley at Bryncrug [6050 0350], but others probably resulted from deposition in temporary lakes or are developed downstream of landslips, as south-west of the Tal-y-llyn landslip.

Between Machynlleth and Penegoes, the top of a broad, gently inclined, terraced fan deposit lies 6 m above the level of the present floodplain of the Dyfi and Dulas valleys. It forms a dry valley which was clearly an important line of east–west drainage before the incision, by the Afon Dulas, of a north–south gorge near Dolguog Hall [7610 0150]. The deposit comprises poorly sorted, imbricated mudstone and sandstone cobbles up to 0.2 m across, in a silty sand matrix. The gravels are locally bedded [7574 0086] and cross-bedded [7623 0053]. Periglacial conditions are indicated by ice-wedges [7525 0060] up to 1.3 m deep.

River terrace deposits and alluvium

The major river terrace deposits and alluvium are confined to the Dysynni, Tal-y-llyn and Dyfi valleys. Most river terraces are less than a few metres high, for example those north of Ty'n-y-bryn [6572 0819] and in the Dyfi valley [8010 0420]. However, higher terraces, up to 6 m above the present floodplain, occur upstream of Abergynolwyn and reflect a thick accumulation prior to deep incision, probably reflecting the capture of the Afon Dysynni from the Tal-y-llyn valley to the Dysynni valley. South of Meriafel uchaf [6841 0764] a meander section of the Afon Dysynni exposes 6 m of imbricated gravels and sands with cross-stratification and channels (Plate 16b). Watson (1965, p. 455) described ice-wedge structures from the upper part of the deposit and its northern margin is overlain by a thick head deposit, derived from the valley side.

The alluvium comprises coarse- to fine-grained gravels with impersistent interbeds of sandy clay, as exposed about Afon Dysynni, north of Llanllwyda [6508 0770]. Downstream, at Bird Rock, the alluvial debris passes into clays and silts. Between Dysefin-uchaf and Llanegryn the alluvial plain is 150–850 m wide, with well-defined abandoned meanders, and locally [6081 0511] the alluvium contains a peat bed, 0.4 m thick. Similar sections of variable lithology occur along the alluvial tract in the Tal-y-llyn valley although there is evidence that, in places, the alluvium is probably lacustrine.

In the Dulas valley, south of Corris, a well-defined alluvial tract, 100–250 m wide, lies close to its confluence with the Dyfi valley and its major alluvial floodplain, up to 1150 m wide. This apparently exceedingly flat tract is crossed by old river courses although these features have been suppressed by successive cultivation. To the south-west of Machynlleth [7306 0039], two narrow river terraces have been distinguished below the level of the current floodplain, about 4.5 m above the normal course of the Afon Dyfi.

Lacustrine deposits

Examination of lacustrine deposits is generally restricted to times of low standing water, to remnant terraces of previous higher stands, or to boreholes. At Tal-y-llyn [7240 1037], the alluvium comprises brown peaty clay, over 0.6 m thick, with faint bedding and abundant rootlets.

More difficult to recognise, particularly in the large valleys, are the deposits of drained lakes. In the deeply excavated valleys, such as Tal-y-llyn, many temporary lakes probably developed behind barriers, such as landslips and alluvial fans. An area of lacustrine alluvium near Tan-y-coed isaf [6550 0526], probably developed behind the large alluvial fan at Dolgoch. The deposit is progessively coarser towards the north-east, in the direction of stream input.

Llyn Gwernan [7030 1590] is a key site for dating the glaciation in the district (Lowe, 1981; Lowe et al., 1988). It contains about 3.5 m of Devensian late-glacial sediments, organic lake mud, silt and clay, and about 10 m of Holocene deposits which on the upstream, western side of the lake have raised the bog above the level of the lake. The pollen and radiocarbon analyses suggest that relatively open grassland had been established prior to about 14 200 BP, possibly the minimum age for Late Devensian ice retreat (Lowe et al., 1988; Campbell and Bowen, 1989).

Peat

Peat, commonly with tree remains, is an important component of the estuarine, lacustrine and river alluvium deposits. The largest expanse occurs upstream of Pont Dysynni, behind the obstructing Bryncrug fan deposit. However, peat also blankets some smoother, upland areas and is preserved in small, poorly drained hollows in the craggy glaciated terrain. A broad swathe of peat, covering about 7.5 km^2, interdigitates with the estuarine alluvium south-east of the Broad Water. At Tywyn [SN 5820 9930], 3–4 m of peat, with a basal layer of in-situ and fallen tree stumps and roots, overlies at least 3.8 m of blue-grey, estuarine clay (Figure 36). A probable rise in sea-level caused inland migration of the storm beach and the basal, tree-bearing layer is now periodically exposed at low tide. The submerged forest was probably contiguous with that at Borth, which is dated at about 4700 years BP (Wilks, 1979; Cave and Hains, 1986).

The source of most of the streams within the district is clearly associated with peat deposits. Typical is the peat-filled depression, on the south side of Mynydd Pencoed [6974 1007], in which a tributary to Afon Cader rises, and which is the catchment of surface drainage from a considerable area of the surrounding hills. Many similar peat deposits occur throughout the district and many have developed by peat growth along a spring line caused by a fault, for example the Ceunant Fault [6400 1015], or a formational boundary, for example the Nod Glas Formation at Foel Grochan [7682 1118].

Probably the largest developments of upland peat occur on the high, deeply incised plateau between the

Dyfi and Tal-y-llyn valleys. The area is densely forested and numerous drainage ditches expose thick peat and overlying soliflucted deposits. At Cribin Fawr [7950 1530] and Maesglasu [8180 1470], the margins of the peat are commonly eroded into hags, up to 3 m thick.

Landslips

Throughout the district there are many small landslips in drift that are largely superficial. Typical of these are the small arcuate rotational slips in till at Waenfach [5856 0492] and Corris Uchaf [7480 0950]. However, the landslip which dams the valley at the south-west end of Tal-y-llyn is a major feature. It comprises mudstones of the Ceiswyn Formation which were detached from a large concave scar on Craig Goch [7140 0840], high on the south side of the valley (see Frontispiece). The landslip forms hummocky ground and is only exposed along the Afon Dysynni and in the area south of the church [7110 0935]. Across the feature there are some discrepancies in cleavage orientations, but overall it is suggested that the slipped strata remained coherent. To the west, the landslip is overlain by flat-topped fan deposits which probably developed immediately after its emplacement. The northern margin of the landslip abuts the valley side but on the south side of the valley it extends up slope into a smear of deeply furrowed, debris-lag deposits which are well featured below the steep scree slopes. Two further rotational slips, which developed after the main collapse, occur above the main slip. The higher of the two, marked by a peat-filled, bench feature [7120 0912], is only a partial collapse and has dropped the land surface by about 40 m. South-east of the church, the lower rotational slip forms a distinctive, flat-topped bench [7110 0905]. There is no sign of recent movement.

Other landslips, also within the Ceiswyn Formation, occur south-west of Abergynolwyn [6725 0640] and at the head of the Tal-y-llyn valley [7500 1350]. Smaller landslips occur within the Ty'r Gawen Mudstones Formation, at Tyn-y-cornel [6315 0765], and on the north side of Craig Cwm-llwyd [6470 1250]. At Tyn-y-cornel, failure occurred between existing, north-west-striking faults. The land surface above the slip is extensively fissured.

NINE

Economic geology

The district lies at the southern edge of the belt of extensive metalliferous mineralisation in the Harlech district (Allen and Jackson, 1985) and north of the Central Wales Mining Field (Jones, 1922). Within the district there is much evidence of trials and small excavations for metalliferous minerals, particularly in the traditionally gold-bearing Clogau Formation, but this mineral exploration met with little success. The synsedimentary manganese deposits of the Hafotty Formation were worked in the vicinity of Barmouth, on the north side of the estuary (Allen and Jackson, 1985), and oolitic ironstones (Fron Newydd Member) were worked mainly in the vicinity of Cross Foxes. The only worked vein deposits, of lead and zinc, occur in Happy Valley and south of Fairbourne.

The main economic exploitation of the geology in the district has been for roofing slate and building stone. Both have been quarried and mined from Tremadoc, upper Ordovician and Silurian strata. Elsewhere, dolerite intrusions and some acid volcanic rocks have been quarried for building stone and hard rock aggregate, but now the only working quarry is at Tonfanau. Throughout the district, the numerous small quarries in scree and head deposits indicate that they have been, and continue to be, an important source of aggregate for hill roads and tracks.

Metalliferous mineralisation

Around the middle of the 19th century metalliferous veins were exploited, and mines developed, in the vicinity of Tywyn and near Fairbourne. In Happy Valley, north-west-striking, vertical mineralised faults have been worked for zinc and lead. At the Tyddyn Briddell mine [SN 6420 9840], three parallel lodes were worked and, reputedly, 100 tons of 'lead ore' was raised before it was closed in 1851 (Bick, 1978). Specimens of quartz-cemented mudstone breccia, with sphalerite, galena and traces of copper, are widespread in the tips. Over 679 tons of lead and zinc ore were recovered from the largest mine in the valley, referred to variously as Caethle or Melin llyn y pair [SN 6170 9920], which was revived in 1851 after earlier exploration in the middle of the 18th century. There, a lode, about 1 m wide, of quartz, sphalerite and galena cemented mudstone breccia was stoped out.

Near Llwyngwril, in the outcrop of the Clogau and Maentwrog formations, a number of adits and small trials [6087 1172 and 6016 1129], probably for gold, were developed, unsuccessfully, in the 1860s. Most are developed on steeply dipping, barren quartz veins. To the east, a north–south lode within dolerite and mudstone was worked underground at Cyfannedd [6276 1235] for lead and silver, mainly between 1842 and 1851. The spoil tips yield a calcite- and chlorite-cemented breccia with sphalerite, chalcopyrite and galena. During the 1870s the property was developed for slate extraction, which caused the large waste tip, but even this activity had ceased before 1887.

To the south of Llynnau Cregennen, small adits and excavations occur along north-east-striking, steeply dipping, mineralised faults within the Cregennen Formation and in the pillow lavas of the Llyn y Gafr Volcanic Formation. The lodes, up to 0.3 m wide, comprise banded jasper and hematite, with common magnetite octahedra.

Although not exploited, the Tonfanau dolerite intrusion contains several north-east-striking zones of ochreous weathered pyritisation, particularly along the contact with the country rock. The minerals, mainly calcite, chalcopyrite and pyrite, are commonly disseminated throughout the rock, but zones of quartz- and calcite-cemented mineralised dolerite breccia are also widespread.

Slate

Slates of variable quality have been quarried from Tremadoc strata and most of the formations above the Aran Volcanic Group. The main development of the industry occurred in the 19th century; Bryn-Eglwys quarry, at Abergynolwyn, and some of the Corris quarries were in production in the 1840s. In 1859, the opening of the Corris–Machynlleth tramroad, known as the Corris Railway, facilitated the transport of the slates to the port at Derwenlas, on the Afon Dyfi. Soon after, in 1864, the Aberdyfi Slate Company enlarged the Bryn-Eglwys operation and used the Tal-y-llyn Railway to transport the slates to Aberdyfi. The quarries produced up to 8000 tonnes per annum in the 1870s and an overall estimated 300 000 tonnes by the time they were closed in 1947 (Holmes and Thomas, 1977).

In the Ceiswyn Formation most of the slate was obtained from the basal 50 m, close to the contact with the underlying, competent volcanic rocks of the Aran Volcanic Group. Quarries occur at Pont y Garth [6342 0704] and Perfeddnant [6300 0536] although the largest excavations are mines and open quarries in Cloddfa Caverns [7977 1597] at Cribin Fawr, in the north-east of the district.

The main extraction in the Corris slate belt occurred above the Nod Glas Formation. The Narrow Vein Formation was the most important economically but the Broad Vein Formation was also intensely excavated in the tract between Abergynolwyn, Corris and Aberllefenni. Farther to the north-east the quality of the slate diminishes as the Broad Vein Formation becomes slightly calcareous.

The compact silty mudstones of the Broad Vein and Narrow Vein formations cleave into slabs and as a result they have been exploited mainly for walls and facings rather than for roofing slates. Most excavations of the Narrow Vein Formation are underground, and, locally, well-developed joints reduce its economic potential. For example, at Godre Fynydd [7660 0990], near Aberllefenni, strongly jointed slate near a major fault is unsuitable for architectural slate and has thus been utilised for small roofing slates. Apart from the small operation at Aberllefenni [7697 1033] (Plate 10a), which produces about 500 tons of finished slate sills and slabs per annum, all activity has ceased. However, the numerous abandoned caverns have considerable potential as tourist attractions.

The Garnedd-wen Formation has been quarried mainly during access to the underlying Narrow Vein Formation, as at Abergynolwyn. The irregular cleavage restricted its use to walls and buildings in the vicinity. The Mottled Mudstone Member and the overlying strata in the Cwmere Formation have been worked for similar use at various places, such as near Pandy in the Dulas Valley [7599 0812] and south of Tarren y Gesail [7203 0576]. Large quarries in the Cwmsymlog Formation [7218 0562], Devil's Bridge Formation [7225 0552] and Borth Mudstones Formation [7672 0610, 7655 0619] were probably developed because of relatively easy access, as the thinly bedded sandstones form poor-quality slabs.

Within the district, there are large reserves of slate waste which have some potential as aggregate. Most emanated from excavations of the Narrow Vein Formation, for example Bryn-Eglwys [6925 0575], Corris [7550 0790], Corris-uchaf [7450 0875], Aberllefenni [7700 1000] and Cwm Ratgoed [7800 1175].

Superficial deposits

Peat has been widely dug for fuel, as near Abergynolwyn [6500 0610] and at the submerged forest at Tywyn [SN 5800 9970], but the thick deposits of the lower Dysynni valley remain largely unexploited. Probable concealed peats in the estuarine alluvium on the northern edge of Tywyn, and in the alluvial deposits of the Tal-y-llyn valley, such as east of Abergynolwyn, create a risk of localised subsidence and site investigation should precede any development.

Within the district, sand and gravel potential is limited; small workings occur within the glaciofluvial sands and gravels at Ynysmaengwyn [5980 0190] and there may be more extensive, concealed deposits at Tonfanau.

REFERENCES

Most of the references listed below are held in the Library of the British Geological Survey at Keyworth, Nottingham. Copies of the references can be purchased subject to the current copyright legislation.

ALLEN, P M, COOPER, D C, and SMITH, I F. 1979. Mineral exploration of the Harlech Dome, North Wales. *Mineral Reconnaissance Report of the Institute of Geological Sciences*, No. 29.

— and JACKSON, A A. 1978. Bryn-teg Borehole, North Wales. *Bulletin of the Geological Survey of Great Britain*, No. 61.

— — 1985. Geology of the country around Harlech. *Memoir of the British Geological Survey*, Sheet 135 (England and Wales).

— — and RUSHTON, A W A. 1981. The stratigraphy of the Mawddach Group in the Cambrian succession of North Wales. *Proceedings of the Yorkshire Geological Society*, Vol. 43, 295–329.

ANDREW, A R. 1910. The geology of the Dolgelly Gold Belt, North Wales. *Geological Magazine*, Vol. 7, 159–171, 201–211, 261–271.

BAKER, S J. 1981. The graptolite biostratigraphy of a Llandovery outlier near Llanystumdwy, Gwynedd, North Wales. *Geological Magazine*, Vol. 118, 355–365.

BASSETT, D A. 1969. Some of the major structures of early Palaeozoic age in Wales and the Welsh Borderland: an historical essay. 67–116 in *The Precambrian and Lower Palaeozoic rocks of Wales*. WOOD, A (editor). (Cardiff: University of Wales Press.)

— WHITTINGTON, H B, and WILLIAMS, A. 1966. The stratigraphy of the Bala District, Merionethshire. *Quarterly Journal of the Geological Society of London*, Vol. 122, 219–271.

BATES, D E B. 1972. The stratigraphy of the Ordovician rocks of Anglesey. *Geological Journal*, Vol. 8, 29–58.

BECKLY, A J. 1987. Basin development in North Wales during the Arenig. 19–30 *in* Sedimentation and tectonics of the Welsh Basin. FITCHES, W R, and WOODCOCK, N H (editors). *Geological Journal*, Vol. 22 (thematic issue).

BELT, T. 1867–68. On the 'Lingula Flags' or 'Ffestiniog Group' of the Dolgelly district. *Geological Magazine*, Vol. 4, (1867), 493–495, 53–543, Vol. 5 (1868), 5–11.

BENNETT, M A. 1987. Genesis and diagenesis of the Cambrian manganese deposits, Harlech, North Wales. 7–18 *in* Sedimentation and tectonics of the Welsh Basin. FITCHES, W R, and WOODCOCK, N H (editors). *Geological Journal*, Vol. 22 (thematic issue).

BICK, D E. 1978. *The old metal mines of Mid-Wales, Part 5: Aberdovey, Dinas Mawddwy and Llangynog.* (Newent: The Pound House.)

BLUNDELL, D J, DAVEY, F J, and GRAVES, L J. 1971. Geophysical surveys over the South Irish Sea and Nymphe Bank. *Journal of the Geological Society of London*, Vol. 127, 339–375.

— GRIFFITHS, D H, and KING, R F. 1969. Geophysical investigations of buried river valleys around Cardigan Bay. *Geological Journal*, Vol. 6, 161–181.

— KING, R F, and WILSON, C D V. 1964. Seismic investigations of the rocks beneath the northern part of Cardigan Bay, Wales. *Quarterly Journal of the Geological Society of London*, Vol. 120, 35–50.

BORRADAILE, G J. 1978. Transected folds: A study illustrated with examples from Canada and Scotland. *Bulletin of the Geological Society of America*, Vol. 89, 481–493.

BRACEGIRDLE, D R. 1974. Structure and stratigraphy of the Towyn–Abergynolwyn area of southern Merionethshire in relation to the Bala Fault. Unpublished MSc thesis, University College of Wales, Aberystwyth.

BRENCHLEY, P J. 1984. Late Ordovician extinctions and their relationship to the Gondwana glaciation. 291–315 in *Fossils and climate.* BRENCHLEY, P J (editor). (Chichester: Wiley.)

— and NEWALL, G. 1984. Late Ordovician environmental changes and their effect on faunas. 65–79 *in* Aspects of the Ordovician System. BRUTON, D L (editor). *Palaeontological Contributions from the University of Oslo*, No. 295. Universitetsforlaget.

BRIDGES, P H. 1975. The transgression of a hard substrate shelf: the Llandovery (Lower Silurian) of the Welsh Borderland. *Journal of Sedimentary Petrology*, Vol. 45, 79–94.

BRITISH GEOLOGICAL SURVEY. 1986. Bala. England and Wales Sheet 136. Solid. 1:50 000. (Southampton: Ordnance Survey for British Geological Survey.)

BRÜCK, P M, COLTHURST, J R J, FEELY, M, GARDINER, P R R, PENNEY, S R, REEVES, T J, SHANNON, P M, SMITH, D G, and VANGUESTAINE, M. 1979. South East Ireland: Lower Palaeozoic stratigraphy and depositional history. 533–544 *in* The Caledonides of the British Isles—reviewed. HARRIS A L, HOLLAND, C H, and LEAKE, B E (editors). *Special Publication of the Geological Society of London*, No. 8.

BULLERWELL, W, and MCQUILLIN, R. 1969. Preliminary report on a seismic reflection survey in the Southern Irish Sea, July 1968. *Report of the Institute of Geological Sciences*, Vol. 69, No. 2.

BULMAN, O M B, and RUSHTON, A W A. 1973. Tremadoc faunas from boreholes in Central England. *Bulletin of the Geological Survey of Great Britain*, Vol. 43, 1–40.

CAMPBELL, S, and BOWEN, D Q. 1989. *Quaternary of Wales.* Nature Conservancy Council: Geological Conservation Review Series A4.1. (Peterborough: Nature Conservancy Council.)

CAMPBELL, S D G. 1983. The geology of an area between Bala and Betws y Coed, North Wales. Unpublished PhD thesis, University of Cambridge.

CARRUTHERS, R M, FLETCHER, C J N, MCDONALD, A J W, and EVANS, R B. 1992. Some constraints on the form of the Welsh Basin from regional gravity and magnetic data, with particular reference to Central Wales. *Geological Magazine,* Vol. 129, 515–522.

CAVE, R. 1965. The Nod Glas sediments of Caradoc age in North Wales. *Geological Journal*, Vol. 4, 279–298.

— 1975. Glan-fred Borehole, Dol-y-bont. p.7 *in* IGS Boreholes 1974. *Report of the Institute of Geological Sciences*, Vol. 75, No. 7.

— 1979. Sedimentary environments of the basinal Llandovery of mid-Wales. 517–526 *in* The Caledonides of the British Isles, reviewed. Harris A L, Holland C H, and Leake, B E (editors). *Special Publication of the Geological Society of London,* Vol. 8.

— and Hains, B A. 1986. Geology of the country between Aberystwyth and Machynlleth. *Memoir of the British Geological Survey,* Sheet 163 (England and Wales).

— and Price, D. 1978. The Ashgill Series near Welshpool, North Wales. *Geological Magazine,* Vol. 115, 183–194.

— and Scott, M S. 1990. Geological notes and local details for 1:10 000 sheet SH 70 SE (Llanwrin). *British Geological Survey, Technical Report,* WA/90/47.

Cho, M, Liou, J G, and Maruyama, S. 1986. Transition from the zeolite to prehnite-pumpellyite facies in the Karmutsen Metabasites, Vancouver Island, British Columbia. *Journal of Petrology,* Vol. 27, 467–494.

Cocks, L R M, Woodcock, N H, Lane, P D, Rickards, R B, and Temple, J T. 1984. The Llandovery Series of the type area. *Bulletin of the British Museum of Natural History (Geology),* Vol. 38, 131–182.

Cooper, A H, and Molyneux, S G. 1990. The age and correlation of Skiddaw Group (early Ordovician) sediments in the Cross Fell inlier (northern England). *Geological Magazine,* Vol. 127, 147–157.

Cope, J C W, Ingham, J J, and Rawson, P F (editors). 1992. Atlas of palaeogeography and lithofacies. *Memoir of the Geological Society of London,* No. 13.

Cowie, J W, Rushton, A W A, and Stubblefield, C J. 1972. A correlation of Cambrian rocks in the British Isles. *Special Report of the Geological Society of London,* No. 2.

Cox, A H. 1925. The geology of the Cader Idris range (Merioneth). *Quarterly Journal of the Geological Society of London,* Vol. 81, 539–594.

— and Wells, A K. 1921. The Lower Palaeozoic rocks of the Arthog–Dolgelly district (Merionethshire). *Quarterly Journal of the Geological Society of London,* Vol. 76, 254–324.

— — 1927. The geology of the Dolgelly district, Merionethshire. *Proceedings of the Geologists' Association,* Vol. 38, 265–318.

Craig, J. 1985. Tectonic evolution of the area between Borth and Cardigan, Dyfed, West Wales. Unpublished PhD thesis, University College of Wales, Aberystwyth.

Crimes, T P. 1970. A facies analysis of the Cambrian of Wales. *Palaeogeography, Palaeoclimatology and Palaeoecology,* Vol. 7, 113–170.

— and Crossley, J D. 1980. Inter-turbidite bottom current orientation from trace fossils, with an example from the Silurian flysch of Wales. *Journal of Sedimentary Petrology,* Vol. 50, 821–830.

Curtis, C D. 1977. Sedimentary geochemistry: environments and processes dominated by involvement in an aqueous phase. *Philosophical Transactions of the Royal Society,* Vol. A286, 353–371.

— 1980. Diagenetic alteration in black shales. *Journal of the Geological Society of London,* Vol. 137, 189–194.

— and Spears, D A. 1968. The formation of sedimentary iron minerals. *Economic Geology,* Vol. 63, 257–270.

Davies, D C. 1878. Treatise on slate and slate quarrying. (London: Crosby, Lockwood and Son.)

Davies, J R, Fletcher, C J N, Waters, R A, Wilson, D, Woodhall, D G, and Zalasiewicz, J A. in press. Geology of the country around Llanilar and Rhayader. *Memoir of the British Geological Survey,* Sheets 178 and 179 (England and Wales).

Davies, R G. 1956. An investigation of the Cadair Idris granophyre and its associated rocks. Unpublished PhD thesis, University College of Wales, Aberystwyth.

— 1959. The Cadair Idris granophyre and its associated rocks. *Quarterly Journal of the Geological Society of London,* Vol. 115, 189–216.

Dean, W T. 1963. The Ordovician trilobite faunas of south Shropshire. Parts III and IV. *Bulletin of the British Museum of Natural History (Geology),* Vol. 8, 213–254, Vol. 9, 1–18.

Dimberline, A J. 1986. Electron microscope and microprobe analysis of chlorite-mica stacks in the Wenlock turbidites, mid-Wales, UK. *Geological Magazine,* Vol. 123, 299–306.

— and Whittington, R J. 1987. The geology of Cardigan Bay. *Proceedings of the Geologists' Association,* Vol. 98, 331–353.

Dobson, M R, Evans, W E, and Whittington, R. 1973. The geology of the south Irish Sea. *Report of the Institute of Geological Sciences,* Vol. 73, No. 11.

Dunkley, P N. 1978. The geology of the south western part of the Aran Range, Merioneth, with particular reference to the igneous history. Unpublished PhD thesis, University College of Wales, Aberystwyth.

— 1979. Ordovician volcanicity of the SE Harlech Dome. 597–601 *in* The Caledonides of the British Isles—reviewed. Harris, A L, Holland, C H, and Leake, B E (editors). *Special Publication of the Geological Society of London.* No. 8.

Ekdale, A A, Muller, L N, and Novak, M T. 1984. Quantitative ichnology of modern pelagic deposits in the abyssal Atlantic. *Palaeogeography, Palaeoclimatology, Palaeoecology,* Vol. 45, 189–223.

Elles, G L. 1922. The Bala country: its structure and rock succession. *Quarterly Journal of the Geological Society of London,* Vol. 78, 132–175.

Fearnsides, W G. 1905. On the geology of Arenig Fawr and Moel Llyfnant. *Quarterly Journal of the Geological Society of London,* Vol. 61, 608–637.

— 1910. Tremadoc slates and associated rocks of south-east Caernarvonshire. *Quarterly Journal of the Geological Society of London,* Vol. 66, 142–188.

Fisher, R V, and Schminke, H -U. 1984. *Pyroclastic rocks.* (Berlin, Heidelberg, New York, Tokyo: Springer Verlag.)

Fitches, W R, and Campbell, S D G. 1987. Tectonic evolution of the Bala Lineament in the Welsh Basin. *Geological Journal,* Vol. 22 (Thematic Issue), 131–153.

Fortey, R A, Beckly, A J, and Rushton, A W A. 1990. The international correlation of the Llanvirn Series, Ordovician System. *Newsletters in Stratigraphy,* Vol. 22, 119–142.

Francis, E H, and Howells, M F. 1973. Transgressive welded ash-flow tuffs among the Ordovician sediments of NE Snowdonia, North Wales. *Journal of the Geological Society of London,* Vol. 129, 621–641.

Geological Survey of Great Britain. 1965. Aeromagnetic map of Great Britain, Sheet 2, England and Wales (south of National Grid line 500 km N). 1:625 000. (Chessington, Surrey: Ordnance Survey for Geological Survey of Great Britain.)

Griffiths, D H, and Gibb, R A. 1965. Bouguer gravity anomalies in Wales. *Geological Journal,* Vol. 4, 335–342.

Hallimond, A F. 1925. Iron ores: bedded ores of England and Wales. Petrography and chemistry. *Special Report on the Mineral Reserves of Great Britain, Memoir of the Geological Survey,* Vol. 25.

HAMBLIN, A P, and WALKER, R G. 1979. Storm dominated shallow marine deposits: the Fernie-Kootenay (Jurassic) transition, southern Rocky Mountains. *Canadian Journal of Earth Sciences,* Vol. 16, 1673–1690.

HARKNESS, R, and HICKS, H. 1871. On the ancient rocks of the St. David's Promontory, South Wales, and their fossil contents. *Quarterly Journal of the Geological Society of London,* Vol. 27, 384–404.

HICKS, H. 1881. The classification of the Eozoic and Lower Palaeozoic rocks of the British Isles. *Popular Science Review,* Vol. 5, 289–308.

HOLMES, A T, and THOMAS, R. 1977. Quarry tracks, village ways. A descriptive history of Bryneglwys Quarry and Abergynolwyn Village. (Tywyn: Talyllyn Railway Company.)

HOWELLS, M F, REEDMAN, A J, and CAMPBELL, S D G. 1991. *Ordovician (Caradoc) marginal basin volcanism in Snowdonia (north-west Wales).* (London: HMSO for the British Geological Survey.)

HUGHES, R A. 1989. Llandeilo and Caradoc graptolites of the Builth and Shelve inliers. *Monograph of the Palaeontographical Society,* No. 577.

INSTITUTE OF GEOLOGICAL SCIENCES. 1971. *Annual report for 1971.* (London: Institute of Geological Sciences.)

— 1973. Central Wales mining field, Scale 1:100 000.

— 1982. Harlech. England and Wales Sheet 135 and part of 149. Solid. 1:50 000. (Southampton: Ordnance Survey for British Geological Survey.)

JAMES, D M D. 1968. Sedimentary studies in the Bala of central Wales. Unpublished PhD thesis, University of Wales, Swansea.

— 1973. The Garnedd-wen Formation (Ashgillian) of the Towyn–Abergynolwyn district, Merionethshire. *Geological Magazine,* Vol. 110, 145–152.

— 1985. Relative sea-level movements, palaeo-horizontals and the depositional relationships of Upper Ordovician sediments between Corris and Bala, Mid Wales. *Mercian Geologist,* Vol. 10, 19–26.

— and JAMES, J. 1969. The influence of deep fractures on some areas of Ashgillian–Llandoverian sedimentation in Wales. *Geological Magazine,* Vol. 106, 562–582.

JEHU, R M. 1926. The geology of the district around Towyn and Abergynolwyn (Merioneth). *Quarterly Journal of the Geological Society of London,* Vol. 82, 465–489.

JENNINGS, A V, and WILLIAMS, G J. 1891. Manod and the Moelwyns. *Quarterly Journal of the Geological Society of London,* Vol. 47, 368–383.

JOHNSON, T E. 1991. Nomenclature and geometric classification of cleavage-transected folds. *Journal of Structural Geology,* Vol. 13, 261–274.

JONES, B. 1933. The geology of the Fairbourne–Llwyngwril district, Merionethshire. *Quarterly Journal of the Geological Society of London,* Vol. 89, 145–171.

JONES, O T. 1909. The Hartfell–Valentian succession in the district around Plynlimon and Pont Erwyd (North Cardiganshire). *Quarterly Journal of the Geological Society of London,* Vol. 65, 463–537.

— 1922. Lead and zinc. The mining district of north Cardiganshire and west Montgomeryshire. *Special Reports on the Mineral Resources of Great Britain. Memoir of the Geological Survey of Great Britain,* Vol. 20.

— 1956. The geological evolution of Wales and the adjacent regions. *Quarterly Journal of the Geological Society of London,* Vol. 111, 323–351.

— and PUGH, W J. 1916. The geology of the district around Machynlleth and the Llyfnant Valley. *Quarterly Journal of the Geological Society of London,* Vol. 71, 343–385.

— — 1935. The geology of the districts around Machynlleth and Aberystwyth. *Proceedings of the Geologists' Association,* Vol. 46, 247–300.

KIMBERLEY, M M. 1974. Origin of iron ore by diagenetic replacement of calcareous oolite. *Nature, London,* Vol. 250, 319–320.

— 1979. Origin of oolitic iron formations. *Journal of Sedimentary Petrology,* Vol. 49, 111–131.

KOKELAAR, B P. 1979. Tremadoc to Llanvirn volcanism on the south east side of the Harlech dome (Rhobell Fawr), North Wales. 591–596 *in* The Caledonides of the British Isles—reviewed. HARRIS, A L, HOLLAND, C H, and LEAKE, B E (editors). *Special Publication of the Geological Society of London,* No. 8.

— 1982. Fluidisation of wet sediments during the emplacement and cooling of various igneous bodies. *Journal of the Geological Society of London,* Vol. 139, 21–33.

— 1986. Petrology and geochemistry of the Rhobell Volcanic Complex: amphibole-dominated fractionation at an Ordovician arc volcano in Wales. *Journal of Petrology,* Vol. 27, 887–914.

— 1988. Tectonic controls of Ordovician arc and marginal basin volcanism in Wales. *Journal of the Geological Society of London,* Vol. 145, 759–775.

— FITCH, F J, and HOOKER, P J. 1982. A new K-Ar age from uppermost Tremadoc rocks of North Wales. *Geological Magazine,* Vol. 119, 207–211.

— HOWELLS, M F, BEVINS, R E, ROACH, R A, and DUNKLEY, P N. 1984. The Ordovician marginal basin of Wales. 245–269. *in* Marginal basin geology; volcanic and associated sedimentary and tectonic processes in modern and ancient marginal basins. KOKELAAR, B P, and HOWELLS, M F (editors). *Special Publication of the Geological Society of London,* No. 16.

LAKE, P, and REYNOLDS, S H. 1896. The Lingula Flags and igneous rocks of the neighbourhood of Dolgelly. *Quarterly Journal of the Geological Society of London,* Vol. 52, 511–522.

— — 1912. The geology of Mynydd-y-Gader, Dolgelly. *Quarterly Journal of the Geological Society of London,* Vol. 68, 345–362.

LAPWORTH, C. 1879. On the tripartite classification of the Lower Palaeozoic rocks. *Geological Magazine,* Vol. 16, 1–15.

LEGGETT, J K. 1980. British Lower Palaeozoic black shales and their palaeo-oceanographic significance. *Journal of the Geological Society of London,* Vol. 137, 139–156.

— MCKERROW, W S, COCKS, L R M, and RICKARDS, R B. 1981. Periodicity in the early Palaeozoic marine realm. *Journal of the Geological Society of London,* Vol. 138, 167–176.

— — and SOPER, N J. 1988. A model for the crustal evolution of southern Scotland. *Tectonics,* Vol. 2, 187–210.

LENG, M J. 1990. Late Ordovician–early Silurian palaeo-environmental analysis in the Tywyn-Corris area of mid-Wales. Unpublished PhD thesis, University College of Wales, Aberystwyth.

LIOU, J G, MARUYANAN, S, and CHO, M. 1985. Phase equilibria and mixed parageneses of metabasites in low-grade metamorphism. *Mineralogical Magazine,* Vol. 49, 321–333.

LOCKLEY, M G. 1980a. The geology of the Llanuwchllyn to Llanymawddwy area, south Gwynedd, North Wales. *Geological Journal*, Vol. 15, 21–41.

— 1980b. Caradoc faunal associations of the area between Bala and Dinas Mawddwy, North Wales. *Bulletin of the British Museum of Natural History (Geology)*, Vol. 33, 165–235.

LOWE, J J, LOWE, S, FOWLER, A J, HEDGES, R E M, and AUSTIN, T J F. 1988. Comparison of accelerator and radiometric radiocarbon measurements obtained from Late Devensian late-glacial lake sediments from Llyn Gwernan, North Wales, UK. *Boreas*, Vol. 17, 355–369.

LOWE, S. 1981. Radiocarbon dating and stratigraphic resolution in Welsh late glacial chronology. *Nature, London*, Vol. 293, 210–212.

LOYDELL, D K. 1991. The biostratigraphy and formational relationships of the Upper Aeronian and Lower Telychian (Llandovery, Silurian) formations of western mid-Wales. *Geological Journal*, Vol. 26, 209–244.

LYNAS, B D T. 1973. The Cambrian and Ordovician rocks of the Migneint area, North Wales. *Journal of the Geological Society of London*, Vol. 129, 481–503.

MARTIN, B A, HOWELLS, M F, and REEDMAN, A J. 1981. A geological reconnaissance study of the Dyfi valley region, Gwynedd/Powys, Wales. *Institute of Geological Sciences/ Environmental Protection Unit*, Report 81–1.

MASSON SMITH, D, HOWELL, P M, and ABERNETHY-CLARK, A B D E. 1974. The National Gravity Reference Net 1973 (NGRN 73). *Professional Papers of the Ordnance Survey*, New Series No. 26.

MATLEY, C A, and WILSON, T S. 1946. The Harlech Dome, north of the Barmouth estuary. *Quarterly Journal of the Geological Society of London*, Vol. 102, 1–40.

MERRIMAN, R J, BEVINS, R E, and BALL, T K. 1986. Petrological and geochemical variations within the Tal-y-Fan Intrusion: a study of element mobility during low-grade metamorphism with implications for petrotectonic modelling. *Journal of Petrology*, Vol. 27, 1409–1436.

MILODOWSKI, A E, and ZALASIEWICZ, J A. 1991a. The origin and sedimentary, diagenetic and metamorphic evolution of chlorite-mica stacks in Llandovery sediments of Central Wales, UK. *Geological Magazine*, Vol. 128, 263–278.

— — 1991b. Redistribution of rare earth elements during diagenesis of turbidite/hemipelagite mudrock sequences of Llandovery age from central Wales. 101–124 *in* Developments in sedimentary provenance studies. MORTON, A C, TODD, S P, and HAUGHTON, P D W (editors). *Special Publication of the Geological Society of London*, No. 57.

MOHR, P A. 1964. Genesis of the Cambrian manganese carbonate rocks of North Wales. *Journal of Sedimentary Petrology*, Vol. 34, 819–829.

MOLYNEUX, S G. 1987. The Arenig Series in South Wales: stratigraphy and palaeontology II. Appendix. Acritarchs and Chitinozoa from the Arenig Series of south-west Wales. *Bulletin of the British Museum of Natural History (Geology)*. Vol. 41, 309–364.

— 1990. Advances and problems in Ordovician palynology of England and Wales. *Journal of the Geological Society of London*, Vol. 147, 615–618.

— and DORNING, K J. 1989. Acritarch dating of latest Tremadoc–earliest Arenig (early Ordovician) sediments in the Camarthen district, south Wales. *Geological Magazine*, Vol. 126, 707–714.

— and RUSHTON, A W A. 1988. The age of the Watch Hill Grits (Ordovician), English Lake District: structural and palaeogeographical implications. *Transactions of the Royal Society of Edinburgh: Earth Sciences*, Vol. 79, 43–69.

NEWELL, N D, PURDEY, E G, and IMBRIE, J. 1960. Bahamian oolite sand. *Journal of Geology*, Vol. 68, 481–497.

NICHOLAS, T C. 1915. Geology of the St. Tudwal's Peninsula (Carnarvonshire). *Quarterly Journal of the Geological Society of London*, Vol. 71, 123–143.

NILSSON, R. 1977. A boring through Middle and Upper Ordovician strata at Koängen in western Scania, southern Sweden. *Sveriges Geologiska Undersökning* (Serie C. Nr. 733), 1–58.

ODIN, G S, and MATTER, A. 1981. De glauconiarum origine. *Sedimentology*, Vol. 28, 611–641.

POWELL, C McA. 1979. A morphological classification of rock cleavages. *Tectonophysics*, Vol. 58, 21–34.

POWELL, D W. 1956. Gravity and magnetic anomalies in North Wales. *Quarterly Journal of the Geological Society of London*, Vol. 111, 375–397.

PRATT, W T. 1990. Tectonics and Caradocian sedimentation in the Tywyn-Corris area of mid-Wales. Unpublished PhD thesis, University College of Wales, Aberystwyth.

— 1991. The Nod Glas décollement of the Welsh Basin. *Geological Magazine*, Vol. 128, 279–282.

— 1992. The use of kink bands to constrain fault displacements: an example from the Bala Lineament, Wales. *Geological Magazine*, Vol. 129, 625–632.

— and FITCHES, W R. 1993. The significance of transected folds along the Bala Lineament, Wales. *Journal of Structural Geology*, Vol. 15, 55–68.

PRICE, D, and MAGOR, P M. 1984. The ecological significance of variation in the generic composition of Rawtheyan (late Ordovician) trilobite faunas from North Wales, UK. *Geological Journal*, Vol. 19, 187–200.

PRICE, N B. 1963. The geochemistry of the Menevian rocks of Wales. Unpublished PhD thesis, University College of Wales, Aberystwyth.

PRIOR, D J. 1987. Syntectonic porphyroblast growth in phyllites: textures and processes. *Journal of Metamorphic Geology*, Vol. 5, 27–39.

PUGH, W J. 1923. The geology of the district around Corris and Aberllefenni, Merionethshire. *Quarterly Journal of the Geological Society of London*, Vol. 79, 508–545.

— 1928. Geology of the district around Dinas Mawddwy. *Quarterly Journal of the Geological Society of London*, Vol. 84, 345–381.

— 1929. Geology of the district between Llanymawddwy and Llanuwchllyn (Merioneth). *Quarterly Journal of the Geological Society of London*, Vol. 85, 242–305.

RAMSAY, A C. 1866. The geology of North Wales. *Memoir of the Geological Survey of Great Britain*, Vol. 3 (1st edition).

— 1881. The geology of North Wales. *Memoir of the Geological Survey of Great Britain*, Vol. 3 (2nd edition).

RAMSAY, J G. 1967. *Folding and fracturing of rocks.* (New York: McGraw-Hill.)

RAST, N. 1969. The relationship between Ordovician structure and volcanicity in Wales. 305–335 in *The Precambrian and Lower Palaeozoic rocks of Wales.* WOOD, A (editor). (Cardiff: University of Wales Press.)

Renard, A. 1878. Sur la structure et la composition mineralogique du coticule et sur ses rapports avec le phyllade oligistifere. *Memoires Courrones de l'Acadamie Royale Belge*, 41. [In French].

Rickards, R B. 1964. The graptolitic mudstones and associated facies in the Silurian strata of the Howgill Fells. *Geological Magazine*, Vol. 101, 435–451.

— 1976. The sequence of Silurian graptolite zones in the British Isles. *Geological Journal*, Vol. 11, 153–188.

— 1989. Exploitation of graptolite cladogenesis in Silurian stratigraphy. 267–274 *in* A global standard for the Silurian System. Holland, C H, and Bassett, M G (editors). *National Museum of Wales Geological Series*, No. 9.

Ridgway, J. 1975. The stratigraphy of Ordovician volcanic rocks around the south and east of the Harlech Dome in Merionethshire, North Wales. *Geological Journal*, Vol. 10, 87–106.

Roberts, B, and Merriman, R J. 1985. The distinction between Caledonian burial and regional metamorphism in metapelites from North Wales: an analysis of isocryst patterns. *Journal of the Geological Society of London*, Vol. 142, 615–624.

— — and Pratt, W T. 1991. The influence of strain, lithology and stratigraphical depth on white mica (illite) crystallinity in mudrocks from the vicinity of the Corris Slate Belt, Wales: implications for the timing of metamorphism in the Welsh Basin. *Geological Magazine*, Vol. 128, 633–645.

Robinson, D, and Bevins, R E. 1986. Incipient metamorphism in the Lower Palaeozoic marginal basin of Wales. *Journal of Metamorphic Geology*, Vol. 4, 101–113.

Rushton, A W A. 1982. The biostratigraphy and correlation of the Merioneth–Tremadoc Series boundary in North Wales. 41–59 *in* The Cambro-Ordovician boundary: sections, fossil distributions and correlations. Bassett, M G, and Dean, W T (editors). *National Museum of Wales Geological Series*, No. 3.

— 1990. Ordovician graptolite biostratigraphy in the Welsh Basin: a review. *Journal of the Geological Society of London*, Vol. 147, 611–614.

— and Molyneux, S G. 1989. The biostratigraphic age of the Ordovician Skiddaw Group in the Black Combe Inlier, English Lake District. *Proceedings of the Yorkshire Geological Society*, Vol. 47, 267–276.

— and Smith, M. 1993. Retrodeformation of fossils — a simple technique. *Palaeontology*, Vol. 36, 927–930.

Sedgwick, A. 1844. On the older Palaeozoic rocks of North Wales. *Quarterly Journal of the Geological Society of London*, Vol. 1, 5–20.

— 1852. On the classification and nomenclature of the Lower Palaeozoic rocks of England and Wales. *Quarterly Journal of the Geological Society of London*, Vol. 8, 136–168.

Smith, A J, and Long, G H. 1969. The Upper Llandovery sediments of Wales and the Welsh Borderland. 237–253 in *The Pre-Cambrian and Lower Palaeozoic rocks of Wales*. Wood, A (editor). (Cardiff: University of Wales Press.)

Smith, R D A. 1987a. Structure and deformation history of the Central Wales synclinorium, northeast Dyfed: implications for long-lived basement structure. 183–198 *in* Sedimentation and tectonics of the Welsh Basin. Fitches, W R and Woodcock, N H (editors). *Geological Journal*, Vol. 22 (thematic issue).

— 1987b. Early diagenetic phosphate cements in a turbidite basin. 141–156 *in* Diagenesis of sedimentary sequences. Marshall, J D (editor). *Special Publication of the Geological Society of London*, No. 36.

Soper, N J, Webb, B C, and Woodcock, N H. 1987. Late Caledonian (Acadian) transpression in north-west England: timing, geometry and geotectonic significance. *Proceedings of the Yorkshire Geological Society*, Vol. 46, 175–192.

Strahan, A, Gibson, W, Cantrill, T C, Sherlock, R L, and Dewey, H. 1920. Pre Carboniferous and Carboniferous bedded ores of England and Wales. *Special Report on the Mineral Reserves of Great Britain. Memoir of the Geological Survey of Great Britain*, Vol. 13.

Swift, D J P, Figueiredo, A G, Freeland, G L, and Oertel, G F. 1983. Hummocky cross stratification and megaripples: a geological double standard. *Journal of Sedimentary Petrology*, Vol. 53, 1295–1317.

Taylor, J H. 1949. Petrology of the Northampton Sand Ironstone Formation. *Memoir of the Geological Survey of Great Britain*.

Taylor, K G. 1990. Berthierine from the non-marine Wealden (Early Cretaceous) sediments of south-east England. *Clay Minerals*, Vol. 25, 391–399.

Temple, J T. 1988. Ordovician–Silurian boundary strata in Wales. 65–71 *in* A global analysis of the Ordovician–Silurian boundary. Cocks, L R M, and Rickards R B (editors). *Bulletin of the British Museum of Natural History (Geology)*, Vol. 43.

— and Cave, R. 1992. Preliminary report on the geochemistry and mineralogy of the Nod Glas and related sediments (Ordovician) of Wales. *Geological Magazine*, Vol. 129, 589–594.

Thomas, A T, Owens, R M, and Rushton, A W A. 1984. Trilobites in British stratigraphy. *Special Report of the Geological Society of London*, No. 16.

Traynor, J-J. 1990. Arenig sedimentation and basin tectonics in the Harlech Dome area (Dolgellau Basin), North Wales. *Geological Magazine*, Vol. 127, 13–30.

Trythall, R J B, Eccles, C, Molyneux, S G, and Taylor, W E G. 1987. Age and controls of ironstone deposition (Ordovician), North Wales. 31–43 *in* Sedimentation and tectonics of the Welsh Basin. Fitches, W R and Woodcock, N H (editors). *Geological Journal*, Vol. 22 (thematic issue).

Turner, R E. 1985. Acritarchs from the type area of the Ordovician Llandeilo Series, South Wales. *Palynology*, Vol. 19, 211–234.

Watson, E. 1962. The glacial morphology of the Tal-y-llyn Valley, Merionethshire. *Transactions of the Institute of British Geographers*, Vol. 30, 15–31.

— 1968. Grèzes litées ou eboulis ordonnés tardiglaciaires dans la région d'Aberystwyth, au centre du Pays de Galles. *Bulletin of the Association of French Geographers*, Nos. 338–339, 16–25. [French].

— 1966. Two nivation cirques near Aberystwyth, Wales. *Biuletyn Peryglacjalny*, Vol. 15, 79–101.

— 1967. The periglacial element in the landscape of the Aberystwyth region. Unpublished PhD thesis, University of Wales, Aberystwyth.

— 1968. The periglacial landscape of the Aberystwyth region. 35–49 *in* Geography at Aberystwyth. Bowen E G, Carter, H, and Taylor, J A (editors). (Cardiff: University of Wales Press.)

— 1970. The Cardigan Bay area. 125–145 *in* The glaciations of Wales and adjoining regions. Lewis, C A (editor). (London: Longman.)

— 1977. Mid and North Wales. INQUA X Congress. Guidebook for excursion C9. (Norwich: Geo Abstracts.)

WAUGH, B. 1970. Petrology, provenance and silica diagenesis of the Penrith Sandstone (Lower Permian) of Northwest England. *Journal of Sedimentary Petrology*, Vol. 40, 1226–1240.

WELLS, A K. 1925. The geology of the Rhobell Fawr District (Merioneth). *Quarterly Journal of the Geological Society of London*, Vol. 81, 463–538.

WILKS, P J. 1979. Mid-Holocene sea-level and sedimentation interactions in the Dovey estuary area, Wales. *Palaeogeography, Palaeoclimatology, Palaeoecology*, Vol. 26, 17–36.

WILSON, D, DAVIES, J R, WATERS, R A, and ZALASIEWICZ, J A. 1992. A fault-controlled depositional model for the Aberystwyth Grits turbidite system. *Geological Magazine*, Vol. 129, 595–607.

WOODCOCK, N H. 1990. Transpressive Acadian deformation across the Central Wales Lineament. *Journal of Structural Geology*, Vol. 12, 329–337.

— AWAN, M A, JOHNSON, T E, MACKIE, A H, and SMITH, R D A. 1988. Acadian tectonics in Wales during Avalonian/Laurentia convergence. *Tectonics*, Vol. 7, 483–495.

— and SMALLWOOD, S D. 1987. Late Ordovician shallow marine environments due to glacio-eustatic regression: Scrach Formation, Mid-Wales. *Journal of the Geological Society of London*, Vol. 144, 393–400.

WOODLAND, A W. 1939. The petrography and petrology of the Lower Cambrian manganese ore of western Merionethshire. *Quarterly Journal of the Geological Society of London*, Vol. 95, 1–35.

— (editor). 1971. The Llanbedr (Mochras Farm) Borehole. *Report of the Institute of Geological Sciences*, Vol. 71, No. 18.

WOODLAND, B G. 1964. The nature and origin of cone-in-cone structures. *Fieldiana, Geology*, Vol. 13, 187–305.

WOOLLARD, G P. 1979. The new gravity system–changes in international gravity base values and anomaly values. *Geophysics*, Vol. 44, 1352–1366.

ZALASIEWICZ, J A. 1981. Stratigraphy and palaeontology of the Arenig area, North Wales. Unpublished PhD thesis, University of Cambridge.

— 1990. Silurian graptolite biostratigraphy in the Welsh Basin. *Journal of the Geological Society of London*, Vol. 147, 619–622.

— 1992. Graptolite dating of Ordovician vulcanicity in the Arenig area, North Wales. *Geological Journal*, Vol. 27, 379–389.

ZIEGLER, A M, COCKS, L R M, and MCKERROW, W S. 1968. The Llandovery transgression of the Welsh Borderland. *Palaeontology*, Vol. 11, 736–782.

APPENDIX 1

The diagnostic characteristics of the uppermost Ashgill and Llandovery biozones and subzones

Normalograptus ? persculptus Biozone (Ashgill)

Low diversity diplograptid faunas, within which *Normalograptus? persculptus, N ? parvulus* and *N. normalis* are common, have been assigned to this biozone, although some may belong to the succeeding *acuminatus* Biozone.

Parakidograptus acuminatus Biozone

This biozone, characterised by the eponymous graptolite together with *Akidograptus ascensus*, has not been definitely determined within the district, but it was proved by Jones and Pugh (1916), and has been recognised in the Aberystwyth district (Cave and Hains, 1986).

Atavograptus atavus and *Lagarograptus acinaces* biozones

Sparse faunas including *Normalograptus normalis, N. rectangularis, Rhaphidograptus toernquisti* and *Atavograptus* sp. probably belong within these biozones; the more extensive collections of Jones and Pugh (1916) distinguished two zonal assemblages.

Coronograptus cyphus Biozone

The appearance of monograptids of the *revolutus* group is used here to define the base of the *cyphus* Biozone. *Rhaphidograptus toernquisti, 'Orthograptus' mutabilis* and *Metaclimacograptus undulatus* are also present.

Monograptus triangulatus Biozone

The diverse assemblages of this biozone include *Monograptus triangulatus separatus* and *M. t. triangulatus*, together with *M. communis* and *Pristiograptus concinnus.*

Normalograptus ? magnus Biozone

This biozone is characterized by the association of *N ? magnus* and *Monograptus triangulatus fimbriatus.*

Pribylograptus leptotheca Biozone (equivalent to the *Monograptus argenteus* Biozone)

Both zone fossils have been recorded, together with taxa such as *Monograptus denticulatus* and *Coronograptus gregarius.*

Monograptus convolutus Biozone

The diverse assemblages assigned to this interval are characterised by *M. convolutus, M. lobiferus* and *M. limatulus*, together with *Metaclimacograptus undulatus, M. decipiens* and *Rastrites hybridus hybridus. Cephalograptus cometa extrema*, where present, denotes the upper part of the biozone.

Monograptus sedgwickii Biozone

The zone fossil is typically abundant, with *Metaclimacograptus undulatus* and *Monograptus contortus* also being common locally. Rickards (1976) considers that *Lagarograptus tenuis* characterises the lower part of the biozone (Figure 26). Some of the assemblages assigned to the *sedgwickii* Biozone may belong to the overlying *halli* Biozone (see Loydell, 1991), but the presence of the latter biozone has only been proved at the Derwenlas type locality (Figure 26; Jones and Pugh, 1916).

Monograptus turriculatus Biozone

This biozone has been subdivided into seven subzones (Loydell, 1991; Davies et al., in press). Of these, the two oldest, the *runcinatus* and *gemmatus* subzones, have not been recognised in the district. Fossils from the Devil's Bridge, Borth Mudstones and Blaen Myherin Mudstones formations are almost entirely of the *utilis* Subzone, with one collection of possible *renaudi* Subzone age close to the base of the Devil's Bridge Formation (Figure 27). The *Pristiograptus renaudi* Subzone is characterised by the abundance of the subzonal fossil, together with graptolites including *Streptograptus pseudoruncinatus* and *S. plumosus.* The *Monograptus utilis* Subzone contains a diverse assemblage in addition to the subzonal fossil, including *Monograptus marri, M. halli, M. planus, Pristiograptus bjerringus* and *Streptograptus plumosus.*

APPENDIX 2

Telychian graptolite localities

Devil's Bridge Formation

1 Track east of Maesllwyni Farm, [7870 0104]. Specimen numbers: LA 844–851. Fauna: *Monograptus marri* ?, *M.* cf. *turriculatus, Streptograptus pseudoruncinatus, S. storchi* ?, *Pristiograptus* sp. Age: *turriculatus* Biozone, possibly *utilis* Subzone. Remarks: the age tentatively suggested depends on the overlap of *S. pseudoruncinatus* with the possible examples of *S. storchi* and *M. marri.*

2 Track near Bryn-wg Isaf, 5175 m at 080° from Machynlleth railway station, [7956 0230]. Specimen numbers: LA 795–826. Fauna: *Monograptus utilis, M. bjerreskovae, M. planus, Streptograptus barrandei, S. plumosus, Pristiograptus* sp. Age: *turriculatus* Biozone, *utilis* Subzone.

3 Track by Pwll-glas, 4100 m at 074° from Machynlleth railway station, [7840 0247]. Specimen numbers: LA 777–794. Fauna: *Monograptus* cf. *planus, M. marri, M. utilis*?, *M. turriculatus, M. halli* ?, *Streptograptus pseudoruncinatus, Petalolithus tenuis* s.l. Age: *turriculatus* Biozone, probably *utilis* Subzone.

4 Track, 710 m at 276° from Aberffrydlan, near Llanwrin, [7720 0299]. Specimen numbers: LA 740–770. *Streptograptus pseudoruncinatus, Pristiograptus renaudi* (common), *Streptograptus plumosus, Monograptus halli* ?, *M. bjerreskovae* and *Rastrites maximus.* Age: *turriculatus* Biozone, *renaudi* Subzone.

5 Track, 620 m at 309° from Cilgwyn, [7722 0452]. Specimen numbers: LA 651-694. *Petalograptus giganteus, Streptograptus pseudoruncinatus, Pristiograptus nudus* s.l., *Monograptus marri, M. halli, M. planus, M. turriculatus* s.l., *Pristiograptus bjerringus, Normalograptus* ? sp., *Petalolithus conicus* ? Age: *turriculatus* Biozone, probably *utilis* Subzone, but *P. giganteus* hints at link with *renaudi* Subzone.

6 Track, 670 m at 315° from Cilgwyn, [7725 0463]. Specimen numbers: LA 695–714. *Petalolithus giganteus, Monograptus* cf. *halli, M. planus, Pristiograptus bjerringus, P. nudus* s.l., *Monograptus bjerreskovae*?. Age: see above.

7 Track, 1150 m at 006° from Esgairlwyd, [7820 0862]. Specimen numbers: LA 633–641. *Monograptus turriculatus, M.* cf. *bjerreskovae, M. planus, M. marri* ?, *M. utilis* ? (recorded by D K Loydell as *M. becki* sensu Schauer, but specimen not seen), *Pristiograptus* sp. Age: *turriculatus* Biozone, *utilis* Subzone or older.

Borth Mudstones Formation

1 Roadside by bungalow at Glan-Dulas Mawr, 2250 m at 014° from Machynlleth railway station, [7505 0383]. Specimen numbers: LA 715–739. Fauna: *Monograptus halli, M. turriculatus, M. bjerreskovae, M.* cf. *utilis, M.* cf. *planus, Petalolithus* cf. *conicus.* Age: *turriculatus* Biozone, probably *utilis* Subzone.

Blaen Myherin Mudstones Formation

1 Track south-west of Afon Crewi, 5100 m at 105° from Machynlleth railway station, [7940 0000]. Specimen numbers: LA 827-843. Fauna: cf. *Normalograptus* ? *scalaris, Monograptus turriculatus, M. halli* ?. Age: *turriculatus* Biozone, probably *renaudi-utilis* subzones.

2 Cutting at south side of new farm track just west of R. Crewi, 585 m at 307° from new farmhouse, Fedw, Penegoes, [7938 0003]. Specimen numbers: CAV 1572–1586. Fauna: *Monograptus utilis, M. planus, M. marri, M. involutus, Pristiograptus nudus, Streptograptus plumosus.* Age: *turriculatus* Biozone, *utilis* Subzone.

APPENDIX 3

List of boreholes

This list includes the permanent record number, location, number of boreholes given by original owner, total depth(s), and stratigraphic range of boreholes within the district. Copies of these records may be obtained from the British Geological Survey, Kingsley Dunham Centre, Keyworth, Nottingham NG12 5GG at a fixed tariff.

SN 59 NE/1–32 Penllyn Farm [SN 5852 9945 to SN 5820 9938], 5 m maximum, Quaternary deposits only.

SN 79 NW/1 near Derwenlas, A487 improvements [SN 7303 9994], 5 m maximum in 12 boreholes, Quaternary deposits.

SH 50 SE/1 Dysynni Bridge [5993 0389], 8 boreholes, 10.9–17.22 m, Quaternary deposits on Ceiswyn Formation.

SH 50 SE/2 Tonfanau [5600 0376], 112.93 m, an abbreviated log is given below:

	Thickness m	*Depth* m
Quaternary		
Clays, sands, gravels and cored boulders retrieved at intervals between 6.50 and 36.61 m. Clays typically yellow-grey. Cored boulders of dolerite, rhyolite, green-grey sandstone and granite. Chippings of cleaved grey mudstone, dolerite, fine sandstone, acid and basic igneous rock	36.61	36.61
Tertiary?		
Upward-fining sequences, 1–10 m thick, of clay and silty clay, yellowish grey, silt and sand with pebbles, some quartzose, breccia, 'conglomero-breccia' (composed of angular mudstone and rounded quartz clasts) and gravel. Lignite at 37.80 m, in conglomero-breccia at 40.13 m, emphasises laminations in sand at 61.90 and 66.80 m	49.69	86.30
Sand, clayey sand, buff and white, clay with breccia/clasts. Lignite between 87.75 and 90.64 m. Possible fault between 90.5 and 91 m.	9.00	95.30
Sandy clay, pale blue-green, buff green and grey mottled, with dispersed clasts; Clay, pale greenish grey, minor silt, becoming ochreous brown, buff-yellow, and green mottled downwards. Possible fault between 95.50 and 95.80 m.	11.92	107.22
Cambro-Ordovician		
Breccia, grey 'slate', siltstone and sandstone clasts with sand matrix. Sharp contact with overlying Tertiary? yellow clay. (Breccia logged as 'weathered head deposits of Palaeozoics').	0.78	108.00
Sandstone (greywacke), mudstone and silty mudstone grey, cleaved.	1.75	109.75
Mudstone, siltstone and sandstone, inter-laminated with slumped(?)laminated blocks. Disrupted laminations, but decreases with increased thickness of siltstone and sandstone beds. Drilling terminated in cleaved sandstone	3.18	112.93

SH 60 SW/1–5 Bryncrug, near Tywyn [6041 0354 to 6143 0312], 3.5 m maximum, Quaternary deposits only.

SH 61 SW/1 near Fairbourne [6154 1486], 48 m, Quaternary deposits only.

SH 70 SW/1–13 near Machynlleth [7442 0182, 7459 0062], 22.55 m maximum, Quaternary deposits only.

SH 70 SE/3 3.25 km east of Machynlleth [7797 0286], 40 m, Quaternary deposits only.

SH 71 SW/1–45 near Minffordd, A487 improvements [7324 1029 to 7497 1327], 26 m maximum, Quaternary deposits on Craig Cau (e.g. SH71SW/ 21) or Ceiswyn formations (e.g. SH71SW/7).

SH 71 SW/46–47 near Tal-y-llyn [7250 1027], 10.45 m maximum, Quaternary deposits only.

SH 71 SE/1–3 continuation [7504 1330 to 7507 1335] of boreholes SH 71 SW/1–45.

FOSSIL INDEX

GENERAL INDEX

BRITISH GEOLOGICAL SURVEY

Keyworth, Nottingham NG12 5GG
0115-936 3100

Murchison House, West Mains Road, Edinburgh
EH9 3LA 0131-667 1000

London Information Office, Natural History Museum
Earth Galleries, Exhibition Road, London SW7 2DE
0171-589 4090

The full range of Survey publications is available through the Sales Desks at Keyworth and at Murchison House, Edinburgh, and in the BGS London Information Office in the Natural History Museum (Earth Galleries). The adjacent bookshop stocks the more popular books for sale over the counter. Most BGS books and reports can be bought from HMSO and through HMSO agents and retailers. Maps are listed in the BGS Map Catalogue, and can be bought together with books and reports through BGS-approved stockists and agents as well as direct from BGS.

The British Geological Survey carries out the geological survey of Great Britain and Northern Ireland (the latter as an agency service for the government of Northern Ireland), and of the surrounding continental shelf, as well as its basic research projects. It also undertakes programmes of British technical aid in geology in developing countries as arranged by the Overseas Development Administration.

The British Geological Survey is a component body of the Natural Environment Research Council.

HMSO

HMSO publications are available from:

HMSO Publications Centre
(Mail, fax and telephone orders only)
PO Box 276, London SW8 5DT
Telephone orders 0171-873 9090
General enquiries 0171-873 0011
Queuing system in operation for both numbers
Fax orders 0171-873 8200

HMSO Bookshops
49 High Holborn, London WC1V 6HB
(counter service only)
0171-873 0011 Fax 0171-831 1326
68–69 Bull Street, Birmingham B4 6AD
0121-236 9696 Fax 0121-236 9699
33 Wine Street, Bristol BS1 2BQ
0117-9264306 Fax 0117-9294515
9 Princess Street, Manchester M60 8AS
0161-834 7201 Fax 0161-833 0634
16 Arthur Street, Belfast BT1 4GD
01232-238451 Fax 01232-235401
71 Lothian Road, Edinburgh EH3 9AZ
0131-228 4181 Fax 0131-229 2734

HMSO's Accredited Agents
(see Yellow Pages)

And through good booksellers